CW00550292

Advanced
Blowout
& Well
Control

Advanced
Blowout
& Well
Control

Robert D. Grace
With Contributions By
Bob Cudd, Richard S. Carden,
and Jerald L. Shursen

Gulf Publishing Company
Houston, London, Paris, Zurich, Tokyo

Advanced Blowout and Well Control

Copyright © 1994 by Gulf Publishing Company, Houston, Texas. All rights reserved. Printed in the United States of America. This book, or parts thereof, may not be reproduced in any form without permission of the publishers.

Gulf Publishing Company
Book Division
P.O. Box 2608 Houston, Texas 77252-2608

10 9 8 7 6 5 4 3 2 1

Library of Congress Cataloging-in-Publication Data

Grace, Robert D.
    Advanced blowout and well control / Robert D. Grace.
        p. cm.
    Includes bibliographical references and index.
    ISBN 0-88415-260-X
    1. Oil wells—Blowouts. 2. Gas wells—Blowouts.
I. Title.
TN871.2.G688 1994
622'.3382—dc20   94-19597
                                         CIP

CONTENTS

ACKNOWLEDGMENTS

I would like to acknowledge my contributors, Mr. Bob Cudd, President of Cudd Pressure Control, Mr. Jerald L. Shursen, and Mr. Richard Carden. As a close personal friend and associate for more than 20 years, Bob Cudd contributed not only to this writing but much more to the total experience in this work. I once reflected that Bob knew more than anyone about various aspects of well control. I later realized that he knew more than everyone else together. Bob represents a wealth of experience, knowledge, and expertise.

I would like to acknowledge Jerry Shursen for his contributions. As a close personal friend, business partner and associate, Jerry and I worked to develop many of the concepts presented in this book. Rich Carden, a friend and associate since his student days at Montana Tech, worked diligently to contribute to this book and insure the quality.

I want to acknowledge the staff at GSM who have worked diligently and with professional pride to insure the quality of the work. Particularly, I must mention and thank my secretary Glennda Norman and our computer genius Jerry Yerger.

Finally, for his inspiration, I would like to acknowledge my life long friend Preston Moore.

PREFACE

Well control problems are always interesting. The raw power that is released by nature in the form of an oil or gas well blowing out of control is awesome. Certainly, well control is one thing and WILD well control is something else. There will be well control problems and wild wells as long as there are drilling operations anywhere in the world. It just goes with the territory.

The consequences of failure are severe. Even the most simple blowout situation can result in the loss of millions of dollars in equipment and valuable natural resources. These situations can also result in the loss of something much more valuable—human life. Well control problems and blowouts are not particular. They occur in the operations of the very largest companies as well as the very smallest. They occur in the most complex operations such as deep, high-pressure gas wells, and they occur in the most simple shallow operations. Men have lost their lives when things went wrong at surface pressures of 12000 psi and at surface pressures of 15 psi. The potential for well control problems and blowouts is ever present.

Advanced
Blowout
& Well
Control

CHAPTER ONE
EQUIPMENT
IN
WELL CONTROL OPERATIONS

"........ I could see that we were having a blowout!" Gas to the surface at 0940 hours.

0940 TO 1230 HOURS

Natural gas was at the surface on the casing side very shortly after routing the returning wellbore fluid through the degasser. The crew reported that most of the unions and the flex line were leaking. A 3-½ inch hammer union in the line between the manifold and the atmospheric-type "poor-boy" separator was leaking drilling mud and gas badly. The separator was mounted in the end of the first tank. Gas was being blown from around the bottom of the poor-boy separator. At about 1000 hours, the motors on the rig floor began to rev as a result of gas in the air intake. The crew shut down the motors.

At 1030 hours the annular preventer began leaking very badly. The upper pipe rams were closed.

1230 TO 1400 HOURS

Continuing to attempt to circulate the hole with mud and water.

1400 TO 1500 HOURS

The casing pressure continues to increase. The flow from the well is dry gas. The line between the manifold and the degasser is washing out and the leak is becoming more severe. The flow from the well is switched to the panic line. The panic line is leaking from numerous connections. Flow is to both the panic line and the separator.

The gas around the rig ignited at 1510 hours. The fire was higher than the rig. The derrick fell at 1520 hours.

This excerpt is from an actual drilling report. Well control problems are difficult without mechanical problems. With mechanical problems such as those described in this report, an otherwise routine well control problem escalates into a disastrous blowout. It is common that, in areas where kicks are infrequent, contractors and operators become complacent with poorly designed auxiliary systems. Consequently, when well control problems do occur, the support systems are inadequate, mechanical problems compound the situation, and a disaster follows.

Because this dissertation is presented as an Advanced Blowout and Well Control Operations Manual, it is not its purpose to present the routine discussion of blowout preventers and testing procedures. Rather, it is intended to discuss the aspects of the role of equipment in well control, which commonly contribute to the compounding of the problems. The components of the well control system and the more often encountered problems are discussed.

THE STACK

Interestingly, the industry doesn't experience many failures within the blowout preventer stack itself. There was one instance in Wyoming where a blowout preventer failed because of a casting problem. In another case, the 5000 psi annular failed at 7800 psi. In general, the stack components are very good and very reliable.

A problem that is continually observed is that the equipment doesn't function when needed. At a well at Canadian, Texas, the annular preventer had been closed on a blowout, but the accumulator would not maintain pressure. Two men were standing on the rig floor when the accumulator lost pressure and the annular preventer opened unexpectedly. As quick as the annular opened, the floor was engulfed in a fireball. Fortunately, no one was seriously burned. The source of the fire was never determined. The rig had been completely shut down, but the accumulator system should have been in working order.

At another blowout in Arkansas, nothing worked. The accumulator wasn't rigged up properly, the ancient annular wouldn't work and, when the pipe rams were closed, the ram blocks fell off the transport arms. After that, the rams couldn't be opened. It's difficult to believe that this equipment was operated and tested as often as the reports indicated. In another instance the stack was to be tested prior to drilling the productive interval. The reports showed that the stack had been tested to the full working pressure of 5000 psi. After failing the test, it was found that the bolt holes had rusted out in the preventer! These situations are not unique to one particular area of the world. Rather, they are common throughout the oil fields of the world. Operators should test and operate the components of the stack to be confident that they are functioning properly.

Having a remote accumulator away from any other part of the rig is a good idea. At one location, the accumulator was next to the mud pumps. The well pressured up and blew the vibrator hose between the mud pump and the stand pipe. The first thing that burned was the accumulator. A mud cross such as illustrated in Figure 1.1 would have saved that rig. It would have been possible to vent the well through the panic line and pump through the kill line and kill the well.

This chapter pertains to all operations whether they are offshore, onshore, remote, or in the middle of a city. Some peculiarities persist that require special considerations. For example, all the equipment in an offshore operation is confined to a small space. However, it is important to remember that a well is deep and dumb and doesn't know where it is or that there is some quantity of water below the rig floor. Therefore, when sacrifices are made and compromises are accepted due to self-imposed space limitations, serious consequences can result.

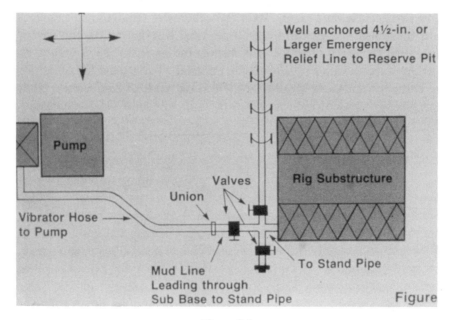

Figure 1.1

THE CHOKE LINE

Many well control problems begin in the choke line or downstream of the choke line. It is unusual to find a rig without the potential for a serious problem between the blowout preventor (BOP) stack and the end of the flare lines. In order to appreciate how a choke line must be constructed, it is necessary to understand that, in a well control situation, solids-laden fluids can be extremely abrasive. The biggest blowout in the history of the state of Texas occurred at the Apache Key 1-11 in Wheeler County. The Apache Key was a cased hole waiting on a pipeline connection when the wellhead blew off. After the well was capped, it began to crater. All vent lines were opened in an attempt to relieve the pressure. As illustrated in Figure 1.2, the 45-degree turns cut out completely. In addition, a close look at Figure 1.2 shows that the 7-1/16-inch-by-10000-psi valve body to the vent line cut out.

At another point in the control operation at the Key, a 20-inch-by-10000-psi stack was being rigged up to cap and vent the well. Due to the size and weight, the stack had to be rigged up in sections. While bringing the second section into place, the crew noticed that the bolts in the first section were loose. The first section was removed and examined. As illustrated in Figure 1.3, the casing head was cut out beyond the ring groove. The inside of all the equipment in the stack and the flow lines had to be protected with a special stellite material. Toward the end of the well control operations, an average of 2,000 cubic yards of particulate material was being removed each month – from a completed well waiting on a pipeline when it blew out.

Figure 1.2

Figure 1.3

It is unusual for dry gas to erode. A production well blowout in North Africa was producing approximately 200 mmscfpd. The dry gas eroded through the tree just as the well was being killed. It had been blowing out for approximately four weeks. Production lines and production chokes which were being used in the well control effort were also severely eroded.

Add a small quantity of carbon dioxide and water and the results can be catastrophic. At a large blowout in East Texas subjected to the described conditions, the blowout preventer was removed after the well was killed. The body of the BOP was almost completely corroded and eroded away.

Unfortunately, the industry has no guidelines for abrasion in the choke line system. Erosion in production equipment is well defined by API RP 14E. Although production equipment is designed for extended life and blowout systems are designed for extreme conditions over short periods of time, the API RP 14E offers insight into the problems and variables associated with the erosion of equipment under blowout conditions. This Recommended Practice relates a critical velocity to the density of the fluid being produced. The equations given by the API are as follows:

$$V_e = \frac{c}{\rho^{\frac{1}{2}}}$$ (1.1)

$$\rho = \frac{12409 S_l P + 2.7 R S_g P}{198.7 P + RTz}$$ (1.2)

$$A = \frac{9.35 + \dfrac{zRT}{21.25P}}{V_e}$$ (1.3)

Where:

V_e	= Fluid erosional velocity, ft/sec
c	= Empirical constant
	= 125 for non-continuous service
	= 100 for continuous service
ρ	= Gas/liquid mixture density at operating temperature, lbs/ft^3
P	= Operating pressure, psia
S_l	= Average liquid specific gravity
R	= Gas/liquid ratio, ft^3/bbl at standard conditions
T	= Operating temperature, °Rankine
S_g	= Gas specific gravity
z	= Gas compressibility factor
A	= Minimum cross-sectional flow area required, in^2/1000 bbls/day

Equations 1.1, 1.2 and 1.3 have been used to construct Figure 1.4, which has been reproduced from API RP 14E and offers insight into the factors effecting erosion. Because the velocity of a compressible fluid increases with decreasing pressure, it is assumed that the area required to avoid erosional velocities increases exponentially with

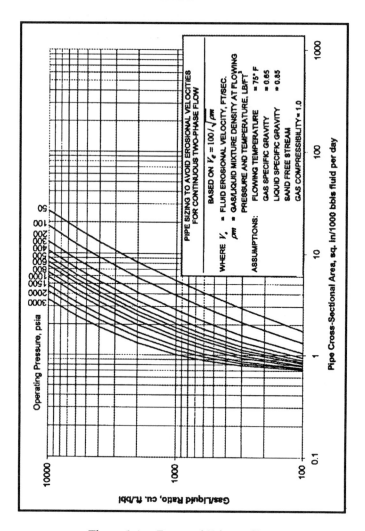

Figure 1.4 - Erosional Velocity Chart

decreasing pressure. However, it is interesting that pursuant to Figure 1.4 and Equations 1.1 through 1.3, a high gas/liquid ratio flow is more erosive than a low gas/liquid ratio flow. One major variable is the hardness of the steel in the component subjected to the blowout condition. A rule of thumb is that under most conditions dry gas does not erode steels harder than N-80 grade.

The presence of solids causes the system to become virtually unpredictable.   Oil-field service companies specializing in fracture stimulation as well as those involved in slurry pipelines are very familiar with the erosional effects of solids in the presence of only liquids. Testing of surface facilities indicates that discharge lines, manifolds and swivel joints containing elbows and short radius bends will remain intact for up to six months at a velocity of approximately 40 feet per second even at pressures up to 15000 psi. Further tests have shown that, in addition to velocity, abrasion is governed by the impingement angle or angle of impact of the slurry solids as well as the strength and ductility of the pipe and the hardness of the solids. At an impingement angle of 10 degrees or less, the erosion wear for a hard, brittle material is essentially zero. In these tests, the maximum wear rate occurred when the impingement angle was between 40 and 50 degrees. The wear rate increased when the solids in the slurry were harder than the tubular surface. Sand is slightly harder than steel. Barite is much less abrasive than hematite.

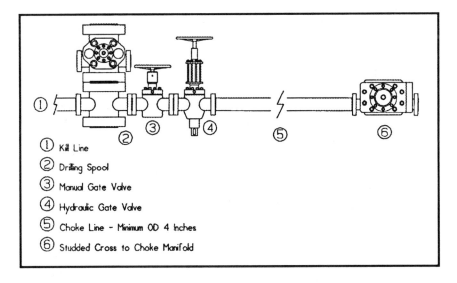

① Kill Line
② Drilling Spool
③ Manual Gate Valve
④ Hydraulic Gate Valve
⑤ Choke Line - Minimum OD 4 Inches
⑥ Studded Cross to Choke Manifold

Figure 1.5 - Choke Line

There is no authority for the erosion and wear rate when solids such as sand and barite are added to gas and drilling mud in a blowout or a well control situation. There can be little doubt that the steels are eroding under most circumstances. API RP 14E merely states that the empirical constant, c, should be reduced if sand production is anticipated. For well control situations, these data dictate that it is imperative that all lines must be straight if at all possible.

A typical choke line is shown in Figure 1.5. As illustrated, two valves are flanged to the drilling spool. There are outlets on the body of the blowout preventers. However, these outlets should not be used on a routine basis since severe body wear and erosion may result. One valve is hydraulically operated while the other is the backup or safety valve. The position of the hydraulic valve is important. Most often it is outboard with a safety valve next to the spool to be used only if the hydraulic valve fails to operate properly. Many operators put the hydraulic valve inboard of the safety valve. Experience has shown that the short interval between the wellbore and the valve can become plugged with drill solids or barite during the normal course of drilling. Therefore, when a problem does occur, the manifold is inoperable due to plugging. The problem has been minimized and often eliminated by placing the hydraulic valve next to the casing spool. The outboard position for the hydraulic valve is the better choice under most circumstances since the inboard valve is always the safety valve. If the hydraulic valve is outboard, it is important that the system be checked and flushed regularly to insure that the choke line is not obstructed with drill solids.

In areas where underbalanced drilling is routine, such as West Texas, drilling with gas influx is normal and the wear on well control equipment can be a serious problem. In these areas, it is not uncommon to have more than one choke line to the manifold. The theory is sound. A backup choke line, in the event that the primary line washes out or is plugged, is an excellent approach. A basic rule in well control is to have redundant systems where a failure in a single piece of equipment does not mean disaster for the operation. However, the second choke line must be as substantial and reliable as the primary choke line. In one instance, the secondary choke line was a 2-inch line from the braden head. The primary choke line failed and the secondary line failed even faster. Since the secondary line was on the braden head with no BOP below, the well blew out under the substructure, caught fire and burned the rig. Therefore, the secondary choke line should come from the kill-line side or from a secondary drilling spool below an additional pipe ram. In addition, it must meet the same specifications for dimension and pressure as the primary choke line.

The choke line from the valves to the choke manifold is a constant problem. This line must be flanged, have a minimum outside diameter of 4 inches and should be STRAIGHT between the stack and the manifold. Any bends, curves or angles are very likely to erode. When that happens, well control becomes very difficult, lost, extremely hazardous or all of the aforementioned.   Just remember,   STRAIGHT   and   no   threaded connections.

If turns in the choke line are required, they should be made with T's and targets as illustrated in Figure 1.6. The targets must be filled with Babot and deep enough to withstand erosion. The direction of flow must be into the target.

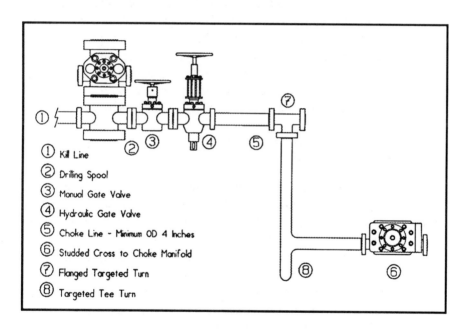

Figure 1.6 - Choke Line with Turns

Figure 1.7 illustrates an improperly constructed choke line. Note that the choke line is bent slightly. In addition, targets are backwards or with the flow from the well. The direction of the targets is a common misunderstanding which has been reported and observed throughout the worldwide industry. These points should be checked on all operations.

Continuous, straight steel lines are the preferred choke lines. Swivel joints should only be used in fracturing and cementing operations and should not be used in a choke line or any well control operation. At a deep, high-pressure sour well in southern Mississippi, a hammer union failed and the rig burned.

Figure 1.7

Finally, the use of hoses has become more popular in recent years. Hoses are quick and convenient to install. However, hoses are recommended only in floating drilling operations which offer no alternative. Further, consider that, in the two most serious well control problems in the North Sea to date, hose failure was the root cause.

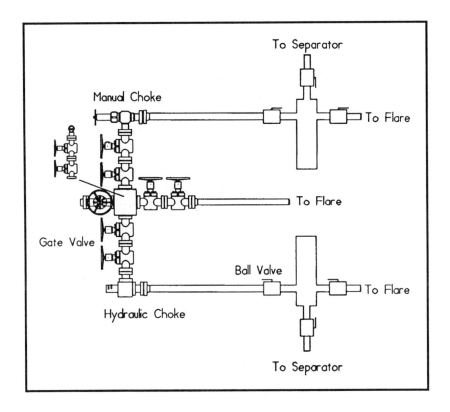

Figure 1.8 *- The Choke Manifold*

Hoses and swivel joints work well on many wells because serious well control problems do not occur on many wells. However, when serious well control problems do occur, solid equipment has better integrity. Swivel joints can be used on the pumping side in kill operations for short periods of time. As of this writing, hose use should be restricted to the suction side of the pumping equipment. Presently, hoses cannot be recommended to replace choke lines. Although the literature is compelling, it is illogical to conclude that rubber is harder than steel.

THE CHOKE MANIFOLD

THE VALVE ARRANGEMENT

A typical choke manifold is illustrated in Figure 1.8. The choke line flanges into a studded cross with two gate valves attached to each outlet of the studded cross. Only the outboard gate valves should be operated routinely. The inboard valves should be used as safety valves in the event the outboard valves cut out. Ball valves are not recommended in these positions because ball valves can be very difficult or impossible to close under adverse conditions. The manifold as illustrated in Figure 1.8 should be considered the minimum requirement for any well.

As the well becomes more complex and the probability of well control problems increases, redundancy in the manifold becomes a necessity. The manifold shown in Figure 1.9 was recently rigged up on a well control problem in the South Texas Gulf Coast. As illustrated in Figure 1.9, there were positions for four chokes on the manifold. Options were good. Either side of the manifold could be the primary side. Since each side was separately manifolded to individual separators, there was redundancy for every system in the manifold. Failure of any single component of the manifold would not jeopardize the operation.

THE PANIC LINE

As illustrated in Figure 1.9, in the center of most land manifolds is a "panic line." This line is usually 4 inches or larger in diameter and goes straight to a flare pit. The idea is that, if the well condition deteriorates to intolerable conditions, the well can be vented to the pit. It is a good idea when properly used and a bad idea when misused. For example, in one instance the rig crew could not get the drilling choke to function properly and, in an effort to relieve the well, the panic line was opened. However, an effort was made to hold back pressure on the well with the valve on the panic line. The valve cut out in less than 30 minutes, making the entire manifold inoperable. There was no choice but to shut in the well and let it blow out underground until the manifold could be repaired.

Valves are made to be open or closed. Chokes are designed to restrict flow. If the panic line is to be used, the line must be fully opened to the flare pit. A better alternative is the manifold system in Figure 1.9. As rigged, the chokes can be opened through the manifolds with some measure of control if desired or necessary.

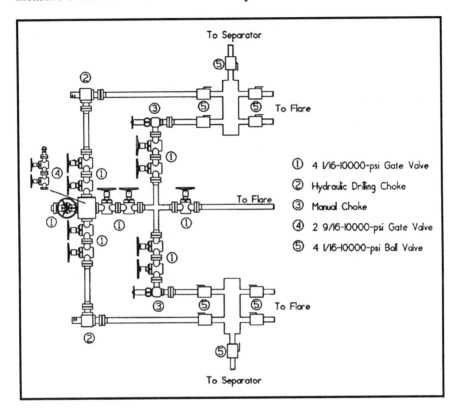

Figure 1.9 - *4-1/16-X-10,000-psi Choke Manifold*

THE DRILLING CHOKE

Outboard of the gate valves are the drilling chokes. The drilling choke is the heart of the well control operation. Well control was not routinely possible prior to the advent of the modern drilling choke. Positive chokes and production chokes were not designed for well control operations and are not tough enough for well control operations. Production chokes and positive chokes should not be included in the choke manifold unless there is a specific production-well testing requirement. Under well control conditions, the best production choke can cut out in a

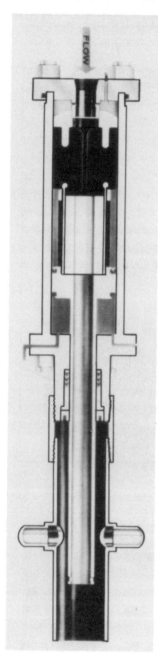

Figure 1.10

few minutes. In addition, if the manifold is spaced for a production choke, a welder will be required to install an additional drilling choke if it is needed. Obviously, it is not good practice to weld on a manifold when well control operations are in progress. If a production choke is used in a manifold system, it is recommended that the system be spaced in order that an additional drilling choke can readily fit. Even with this precaution, there remains the problem of testing newly installed equipment with the well control operation in progress.

In the late 1950s a company known as Drilling Well Control was active controlling kicks with a series of skid-mounted separators. Prior to this technology, it was not possible to hold pressure on the annulus to control the well. It was awkward at best with the separators. The system was normally composed of two to three separators, depending on the anticipated annulus pressures. The annulus pressure was stepped down through the separators in an effort to maintain a specific annular pressure. It represented a significant step forward in technology but pointed to the specific need for a drilling choke that could withstand the erosion resulting from solids-laden multiphase flow.

The first effort was affectionately known as the "horse's a.." choke. The nickname will be apparent as its working mechanism is understood. The choke is illustrated in Figure 1.10. Basically, it was an annular blowout preventer turned on its side. The flow stream entered the rubber bladder. The pressure on the well was maintained by pressuring the back side of the bladder. The solids would cut out the bladder, requiring that the operator pay close attention to the pressures everywhere at the same time, therefore making it an awkward choke to use.

In a short period of time, the technology of drilling chokes advanced dramatically. There are several good drilling chokes available to the industry; however, two are presented in this text. The first really good choke, illustrated in Figure 1.11, was the SWACO Super Choke with two polished plates to restrict flow. One plate is fixed and the other is hydraulically rotated to open or close the choke by rotating the opening in one plate over the opening in the other plate. The choke can also be operated manually in the event of loss of rig air, which is a feature not offered by many other drilling chokes. As is the case with most drilling chokes, the Super Choke is trimmed for sour environments. The polished plates permit testing the choke to full working pressure – a feature not offered by all drilling chokes. The SWACO Super Choke has a history of being very reliable.

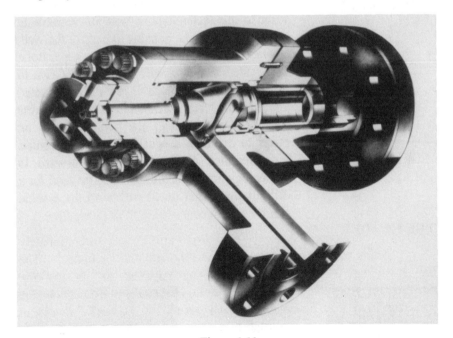

Figure 1.11

The Cameron Drilling Choke, illustrated in Figure 1.12, consists of a tungsten carbide cylinder and a tungsten carbide sleeve. Choking is accomplished by moving the cylinder hydraulically into the sleeve. The

Cameron Drilling Choke cannot be operated manually and cannot be tested to full working pressure. However, it has a long history of reliability under the most adverse conditions.

The drilling choke is the heart of the well control operation. This is not the place to economize. There is no substitute for a good drilling choke. Install it, test it, keep it in good working order and know how it works.

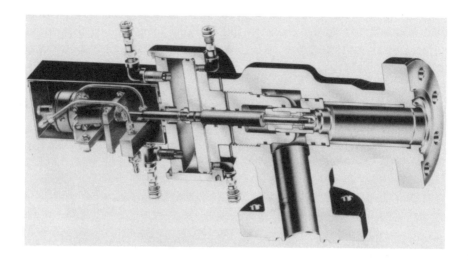

Figure 1.12

THE HEADER

Many equipment failures in well control operations occur between the choke and the separator. With regard to the lines immediately downstream from the chokes, the most common problem is erosion resulting from insufficient size. They should be at least 4 inches in diameter in order to minimize the velocity. Any abrupt change in the diameter will cause an area of more severe erosion. These lines do not have to be high pressure; however, they should have a yield strength of 80000 psi or greater in order that the steel will be harder than most of the particulate in the flow stream and resist erosion. Two-inch lines are too small for any operation and should not be used.

Consistently, a major problem is the header. In most manifold assemblies, the entire system becomes inoperable when the header fails. Consider Figures 1.13 and 1.14 which illustrate a header design common to many operations. This system failed for two reasons. The lines between the choke manifold and the header were 2 inches outside diameter which resulted in excessive velocities and the back side of the header was eroded away by the jetting action of the flow stream as it entered the header and expanded. Obviously, the well is out of control and must be either vented or shut in to blow out underground, depending on the surface pressure and casing design. In most instances, the well is shut in and an underground blowout is the result.

Inadequate design between the chokes and the separators is a problem common to all oil fields. Conceptually, the header is intended to connect the lines from the manifold to the separator. It is convenient to manifold all the lines together into a header and into one line to the separator. The result is that the entire operation depends upon the integrity of that one header and line. In addition, that header and line can be subjected to severe operating conditions. In the past, the solution attempts have been to install a target in the header opposite the incoming line. Another approach has been to make the header much thicker. Many use a combination of the aforementioned.

There are better alternatives. Figure 1.9 illustrates the best alternative. As shown, each side of the manifold has a separate header and separator. With the manifold in Figure 1.9, there is redundancy in all equipment except the choke line. If there is concern about the integrity of the choke line, a second choke line must be installed as previously described.

Figure 1.13

Figure 1.14

Another option, which is illustrated in Figure 1.15, is that the header is above the lines coming from the choke and can be isolated by full-opening ball valves. The downstream valve should be at least 5 diameters from the tee. With this arrangement, the flow can be diverted to the header or to the flare. The wear on the header is significantly reduced because the momentum of the fluid is broken by the tee. However, the wall thickness of the header should be at least 1-½ inches and there should be flanged targets opposite the incoming lines.

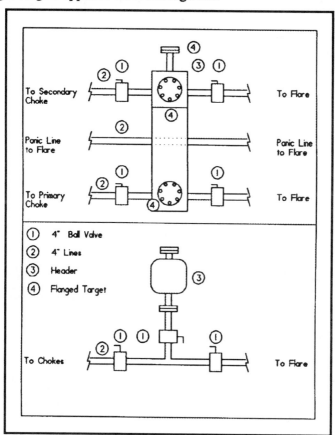

Figure 1.15 - Alternative Choke Manifold

In Figure 1.15, the primary choke could be isolated in the event of a failure by closing the valve in the header. With the valve in the header closed, the secondary choke could be utilized until the failed component in the primary system had been repaired or replaced. The header in

Figure 1.8 would be minimally acceptable provided that a valve was installed in the header to isolate one side of the choke manifold from the other.

THE SEPARATOR

The lines from the header to the separator and the separator itself are areas where problems are common. The problems begin in the line

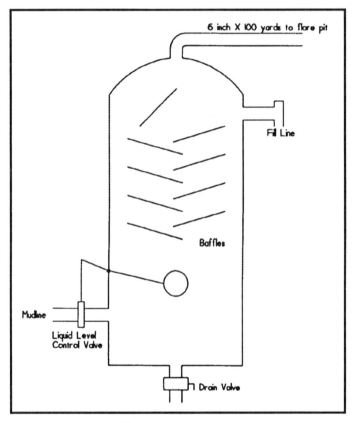

Figure 1.16 - Separator

from the header to the separator. This line must be at least 4 inches in diameter, must be of constant diameter and must be straight or have targeted turns. The targeted turns are more important in this section of the manifold system than in the choke line because the velocities are higher and erosion is a function of velocity. The direction of flow must be into the target. Rotary hoses and other flexible hoses should not be used.

The separator must be BIG. It should be at least 4 feet in diameter and 8 feet tall. In addition, a liquid level control valve is a necessity. Too often, the separator consists of a piece of 30-inch casing with a few baffles and no liquid level control. These "poor-boy" separators are insufficient. As illustrated in the drilling report at the beginning of this chapter, when gas comes to the surface, it is commonly blown from around the bottom. In a difficult situation, gas and mud are emitted through the gas line as well as the liquid line. As a result, gas accumulates on the mud pits and in other areas around the location, creating a very hazardous condition.

The separators on most offshore operations are too small and poorly designed. Gas in the mud line offshore is particularly dangerous since ventilation in the mud room is often limited. A good separator design is illustrated in Figure 1.16.

Finally, the flare line from the separator to the flare boom, top of the derrick or burn pit should be straight and long enough to get the fire a comfortable distance away from the rig floor. Most flare lines are 6 inches in diameter.

THE STABBING VALVE

On all rigs and work-over operations there should be a valve readily accessible on the rig floor which is adapted to the tubulars. Such a valve is routinely called a "stabbing valve." In the event of a kick during a trip, the stabbing valve is installed in the connection on the rig floor and closed to prevent flow through the drill string. Ordinarily, the stabbing valve is a ball valve such as illustrated in Figure 1.17.

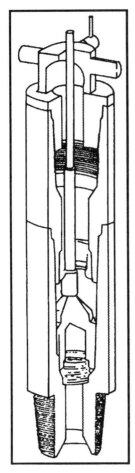

Figure 1.17
Stabbing Valve

The ball valve utilized for a stabbing valve is probably the best alternative available under most circumstances. However, the valve is extremely difficult and problematic to operate under pressure. If the valve is closed and pressure is permitted to build under the valve, it becomes impossible to open the valve without equalizing the pressure. If the pressure below the valve is unknown, the pressure above the valve must be increased in small increments in an effort to equalize and open the valve.

The stabbing valve must have an internal diameter equal to or greater than the tubulars below the valve and the internal diameter of the valve must be documented. Too often, the internal diameter is too small, unknown and undocumented. Therefore, when a well control problem does occur, the first operation is to freeze the stabbing valve and replace it with a valve suitable for the job.

If the flow through the drill string is significant, it may not be possible to close the stabbing valve. In many instances when a flow is observed, the blowout preventers are closed before the stabbing valve. With all of the flow through the drill string, it may not be possible to close the stabbing valve. In that instance, it is necessary to open the blowout preventers, attempt to close the stabbing valve and re-close the blowout preventers. This discussion emphasizes the accessibility of the stabbing valve and the necessity of conducting drills to insure that the stabbing valve can be readily installed and operated.

Finally, when the stabbing valve cannot be installed or closed and shear rams have been included in the stack, there is a tendency to activate the shear rams and sever the drill string. This option should be considered carefully. In most instances, shearing the drill string will result in the loss of the well. The energy released in a deep well, when the drill string is suddenly severed at the surface, is unimaginable and can be several times that required to cause the drill string to destroy itself along with the casing in the well. When that happens, chances for control from the surface are significantly reduced.

CHAPTER TWO
CLASSIC PRESSURE CONTROL PROCEDURES WHILE DRILLING

24 September

0600 - 0630   Service rig.

0630 - 2130   Drilling 12,855 feet to 13,126 feet for 271 feet.

2130 - 2200   Pick up and check for flow.   Well flowing. Shut well in.

2200 - 2300   Pump 35 barrels down drillpipe. Unable to fill drillpipe.

2300 - 2330   Observe well. 1000 psi on casing. 0 psi on drillpipe.

2330 - 0030   Pump 170 barrels down drillpipe using Barrel In - Barrel Out Method. Could not fill drillpipe.   0 psi on drillpipe. 1200 psi on casing at choke   panel. Shut pump off. Check gauge on choke manifold. Casing pressure 4000 psi at choke   manifold.   Choke panel gauge pegged at 1200   psi.   Well out of control.

This drilling report was taken from a recent blowout in the South Texas Gulf Coast. All of the men on the rig had been to well control school and were Minerals Management Service (MMS) certified. After the kick, it was decided to displace the influx using the Barrel In - Barrel

Out   Method.   As a result, control was lost completely and an underground blowout followed.

Prior to 1960, the most common method of well control was known as the Constant Pit Level Method or the Barrel In - Barrel Out Method. However, it was realized that if the influx was anything other than water, this method was catastrophic and classical pressure control procedures were developed. It is incredible that even today there are those in the field who continue to use this antiquated method.

Ironically, there are instances when these methods are appropriate and the classical methods are not. It is equally incredible that in some instances classical procedures are applied to situations which are completely inappropriate. If the actual situation is not approximated by the theoretical models used in the development of the classical procedures, the classical procedures are not appropriate. There is an obvious general lack of understanding. It is the purpose of this chapter to establish firmly the theoretical basis for the classical procedures as well as describe the classical procedures. The application of the theory must be strictly followed in the displacement procedure.

CAUSES OF WELL KICKS AND BLOWOUTS

A kick or blowout may result from one of the following:

1. Mud weight less than formation pore pressure
2. Failure to keep the hole full while tripping
3. Swabbing while tripping
4. Lost circulation
5. Mud cut by gas, water or oil

MUD WEIGHT LESS THAN FORMATION PORE PRESSURE

There has been an emphasis on drilling with mud weights very near to and, in some instances, below formation pore pressures in order to maximize penetration rates. It has been a practice in some areas to take a kick to determine specific pore pressures and reservoir fluid composition. In areas where formation productivity is historically low (roughly less than 1 million standard cubic feet per day without stimulation), operators often drill with mud hydrostatics below the pore pressures.

Mud weight requirements are not always known for certain areas. The ability of the industry to predict formation pressures has improved in recent years and is sophisticated. However, a North Sea wildcat was recently 9 pounds per gallon overbalanced while several development wells in Central America were routinely 2 pounds per gallon underbalanced. Both used the very latest techniques to predict pore pressure while drilling. Many areas are plagued by abnormally pressured, shallow gas sands. Geologic correlation is always subject to interpretation and particularly difficult around salt domes.

FAILURE TO KEEP THE HOLE FULL AND SWABBING WHILE TRIPPING

Failure to keep the hole full and swabbing is one of the most frequent causes of well control problems in drilling. This problem is discussed in depth in Chapter 3.

LOST CIRCULATION

If returns are lost, the resulting loss of hydrostatic pressure will cause any permeable formation containing greater pressures to flow into the wellbore. If the top of the drilling fluid is not visible from the surface, as is the case in many instances, the kick may go unnoticed for some time. This can result in an extremely difficult well control situation.

One defense in these cases is to attempt to fill the hole with water in order that the well may be observed. Usually, if an underground flow is occurring, pressure and hydrocarbons will migrate to the surface within a few hours. In many areas it is forbidden to trip out of the hole without returns to the surface. In any instance, tripping out of the hole without mud at the surface should be done with extreme caution and care, giving consideration to pumping down the annulus while tripping.

MUD CUT

Gas-cut mud has always been considered a warning signal, but not necessarily a serious problem. Calculations demonstrate that severely gas-cut mud causes modest reductions in bottomhole pressures because of the compressibility of the gas. An incompressible fluid such as oil or water can cause more severe reductions in total hydrostatic and has caused serious well control problems when a productive oil or gas zone is present.

INDICATIONS OF A WELL KICK

Early warning signals are as follows:

1.   Sudden increase in drilling rate
2.   Increase in fluid volume at the surface, which is commonly termed a pit level increase or an increase in flow rate
3.   Change in pump pressure
4.   Reduction in drillpipe weight
5.   Gas, oil, or water-cut mud

SUDDEN INCREASE IN DRILLING RATE

Generally, the first indication of a well kick is a sudden increase in drilling rate or a "drilling break," which is interpreted that a porous formation may have been penetrated. Crews should be alerted that, in the potential pay interval, no more than some minimal interval (usually 2 to 5 feet) of any drilling break should be penetrated. This is one of the most important aspects of pressure control. Many multimillion-dollar blowouts could have been avoided by limiting the open interval.

INCREASE IN PIT LEVEL OR FLOW RATE

A variation of bit type may mask a drilling break. In that event, the first warning may be an increase in flow rate or pit level caused by the influx of formation fluids. Depending on the productivity of the formation, the influx may be rapid or virtually imperceptible. Therefore, the influx could be considerable before being noticed. No change in pit level or flow rate should be ignored.

CHANGE IN PUMP PRESSURE

A decrease in pump pressure during an influx is caused by the reduced hydrostatic in the annulus. Most of the time, one of the aforementioned indications will have manifested itself prior to a decrease in pump pressure.

REDUCTION IN DRILLPIPE WEIGHT

The reduction in string weight occurs with a substantial influx from a zone of high productivity. Again, the other indicators will probably have manifested themselves prior to or in conjunction with a reduction in drillpipe weight.

GAS, OIL, OR WATER-CUT MUD

Caution should be exercised when gas, oil, or water-cut mud is observed. Normally, this indicator is accompanied by one of the other indicators if the well is experiencing an influx.

SHUT-IN PROCEDURE

When any of these warning signals are observed, the crew must immediately proceed with the established shut-in procedure. The crew must be thoroughly trained in the procedure to be used and that procedure should be posted in the dog house. It is imperative that the crew be properly trained and react to the situation. Classic pressure control procedures cannot be used successfully to control large kicks. The success of the well control operation depends upon the response of the crew at this most critical phase.

A typical shut-in procedure is as follows:

1. Drill no more than 3 feet of any drilling break.
2. Pick up off bottom, space out and shut off the pump.
3. Check for flow.
4. If flow is observed, shut in the well by opening the choke line, closing the pipe rams and closing the choke, pressure permitting.
5. Record the pit volume increase, drillpipe pressure and annulus pressure. Monitor and record the drillpipe and annular pressures at 15-minute intervals.
6. Close annular preventer; open pipe rams.
7. Prepare to displace the kick.

The number of feet of a drilling break to be drilled prior to shutting in the well can vary from area to area. However, an initial drilling break of 2 to 5 feet is common. The drillpipe should be spaced out to insure that no tool joints are in the blowout preventers. This is especially important on offshore and floating operations. On land, the normal procedure would be to position a tool joint at the connection position above the rotary table to permit easy access for alternate pumps or wire-line operations. The pump should be left on while positioning the drillpipe. The fluid influx is distributed and not in a bubble. In addition, there is less chance of initial bit plugging.

When observing the well for flow, the question is "How long should the well be observed?" The obvious answer is that the well should be observed as long as necessary to satisfy the observer of the condition of the well. Generally, 15 minutes or less are required. If oil muds are being used, the observation period should be increased. If the well is deep, the observation period should be longer than for a shallow well.

If the drilling break is a potentially productive interval but no flow is observed, it may be prudent to circulate bottoms up before continuing drilling in order to monitor and record carefully parameters such as time, strokes, flow rate and pump pressure for indications of potential well control problems. After it is determined that the well is under control, drill another increment of the drilling break and repeat the procedure. Again, there is flexibility in the increment to be drilled. The experience gained from the first increment must be considered. A second increment of 2 to 5 feet is common. Circulating out may not be necessary after each interval even in the productive zone; however, a short circulating period will disperse any influx. Repeat this procedure until the drilling rate returns to normal and the annulus is free of formation fluids.

Whether the annular preventer or the pipe rams are closed first is a matter of choice. The closing time for each blowout preventer must be considered along with the productivity of the formation being penetrated. The objective of the shut-in procedure is to limit the size of the kick. If the annular requires twice as much time to close as the pipe rams and the formation is prolific, the pipe rams may be the better choice. If both blowout preventers close in approximately the same time, the annular is the better choice since it will close on anything.

Shutting in the well by opening the choke, closing the blowout preventers and closing the choke is known as a "soft shut-in." The alternative is known as a "hard shut-in" which is achieved by merely closing the blowout preventer on the closed choke line. The primary argument for the hard shut-in is that it minimizes influx volume, and influx volume is critical to success. The hard shut-in became popular in the early days of well control. Before the advent of modern equipment with remote hydraulic controls, opening choke lines and chokes was time-consuming and could permit significant additional influx. With modern equipment, all hydraulic controls are centrally located and critical valves are hydraulically operated. Therefore, the shut-in is simplified and the time reduced. In addition, blowout preventers, like valves, are made to be open or closed while chokes are made to restrict flow. In some instances, during hard shut-in, the fluid velocity through closing blowout preventers has been sufficient to cut out the preventer before it could be closed effectively.

In the young rocks such as are commonly found in offshore operations, the consequences of exceeding the maximum pressure can be grave in that the blowout can fracture to the surface outside the casing. The blowout then becomes uncontrolled and uncontrollable. Craters can consume jack-up rigs and platforms. The plight of the floating rig can be even more grim due to the loss of buoyancy resulting from gas in the water.

As a final compelling consideration in addition to the others mentioned, historically the most infamous and expensive blowouts in industry history were associated with fracturing to the surface from under surface casing.

It is often argued that fracturing to the surface can be avoided by observing the surface pressure after the well is closed in and opening the well if the pressure becomes too high. Unfortunately, in most instances there is insufficient time to avoid fracturing at the shoe. All things considered, the soft shut-in is the better procedure.

In the event the pressure at the surface reaches the maximum permissible surface pressure, a decision must be made either to let the well blow out underground or to vent the well to the surface. Either approach can result in serious problems. With only surface casing set to a depth of less than 3,600 feet, the best alternative is to open the well and permit the well to flow through the surface equipment. This procedure can result in the erosion of surface equipment. However, more time is made available for rescue operations and repairs to surface equipment. It also simplifies kill operations.

There is no history of a well fracturing to the surface with pipe set below 3,600 feet. Therefore, with pipe set below 3,600 feet, the underground blowout is an alternative. It is argued that an underground flow is not as hazardous as a surface flow in some offshore and land operations. When properly rigged up, flowing the well to the surface under controlled conditions is the preferred alternative. A shut-in well that is blowing out underground is difficult to analyze and often more difficult to control.

The maximum permissible shut-in surface pressure is the lesser of 80 to 90% of the casing burst pressure and the surface pressure required to produce fracturing at the casing shoe. The procedure for determining the maximum permissible shut-in surface pressure is illustrated in Example 2.1:

Example 2.1
   Given:

Surface casing	=	2,000 feet 8 5/8-inch
Internal yield	=	2470 psi
Fracture gradient, F_g	=	0.76 psi/ft
Mud density, ρ	=	9.6 ppg
Mud gradient, ρ_m	=	0.5 psi/ft
Wellbore schematic	=	Figure 2.1

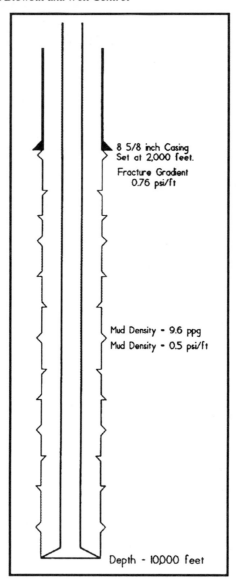

Figure 2.1 - Wellbore Schematic

Required:

Determine the maximum permissible surface pressure on the annulus, assuming that the casing burst is limited to 80% of design specification.

Solution:
80% burst = 0.8 (2470 psi) = **1976 psi**

$$P_f = P_a(Maximum) + \rho_m D_{sc} \qquad (2.1)$$

Where:

P_f	= Fracture pressure, psi
P_a	= Annulus pressure, psi
ρ_m	= Mud gradient, psi/ft
D_{sc}	= Depth to the casing shoe

Therefore:

$$P_a(Maximum) = P_f - \rho_m D_{sc}$$

$$= 0.76 \, (2000) - 0.5 \, (2000)$$

$$= \mathbf{520 \ psi}$$

Therefore, the maximum permissible annular pressure at the surface is 520 psi, which is that pressure which would produce formation fracturing at the casing seat.

Recording the gain in pit volume, drillpipe pressure and annulus pressure initially and over time is very important to controlling the kick. As will be seen in the discussion of special problems in Chapter 4, the surface pressures are critical for determining the condition of the well and the potential success of the well control procedure. Analysis of the gain in the surface volume in consideration of the casing pressure is critical in defining the potential for an underground blowout. In some instances, due to a lack of familiarity with the surface equipment, the crew has failed to shut in the well completely. When the pit volume continued to increase, the oversight was detected and the well shut in. Recording the surface pressures over time is extremely important. Gas migration, which is also discussed in Chapter 4, will cause the surface pressures to increase over time. Failure to recognize the resultant superpressuring can result in the failure of the well control procedure.

These procedures are fundamental to pressure control and represent the most singularly important aspect of pressure control. They are the responsibility of the rig crew and should be practiced and studied until they become as automatic as breathing. The entire operation depends upon the ability of the driller and crew to react to a critical situation. Now, the well is under control and the kill operation can proceed to circulate out the influx.

CIRCULATING OUT THE INFLUX

THEORETICAL CONSIDERATIONS

Gas Expansion

Prior to the early 1960s, an influx was circulated to the surface by keeping the pit level constant. This was also known as the Barrel In - Barrel Out Method. Some insist on using this technique today although it is no more successful now than then. If the influx was mostly liquid, this technique was successful. If the influx was mostly gas, the results were disastrous. When a proponent of the Constant Pit Level Method was asked about the results, he replied, "Oh, we just keep pumping until something breaks!" Invariably, something did break, as illustrated in the drilling report at the beginning of this chapter.

In the late 1950s and early 1960s some began to realize that this Barrel In - Barrel Out technique could not be successful. If the influx was gas, the gas had to be permitted to expand as it came to the surface. The basic relationship of gas behavior is given in Equation 2.2:

$$PV = znRT \qquad (2.2)$$

Where:

P = Pressure, psia
V = Volume, ft^3
z = Compressibility factor

n = Number of moles
R = Units conversion constant
T = Temperature, °Rankine

For the purpose of studying gas under varying conditions, the general relationship can be extended to another form as given in Equation 2.3:

$$\frac{P_1 V_1}{z_1 T_1} = \frac{P_2 V_2}{z_2 T_2}$$ (2.3)

1 - Denotes conditions at any point
2 - Conditions at any point other than point 1

By neglecting changes in temperature, T, and compressibility factor, z, Equation 2.3 can be simplified into Equation 2.4 as follows:

$$P_1 V_1 = P_2 V_2$$ (2.4)

In simple language, Equation 2.4 states that the pressure of a gas multiplied by the volume of the gas is constant. The significance of gas expansion in well control is illustrated by Example 2.2:

Example 2.2
Given:

Wellbore schematic		=	Figure 2.2
Mud density,	ρ	=	9.6 ppg
Mud gradient,	ρ_m	=	0.5 psi/ft
Well depth,	D	=	10,000 feet

Conditions described in Example 2.1

Assume that the wellbore is a closed container.

Assume that 1 cubic foot of gas enters the wellbore.

Assume that gas enters at the bottom of the hole, which is point 1.

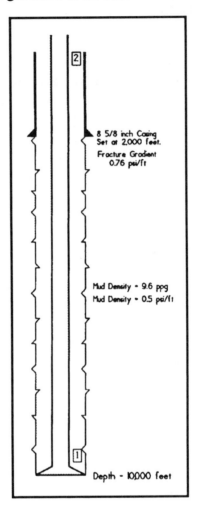

8 5/8 inch Casing
Set at 2,000 feet.
Fracture Gradient
0.76 psi/ft

Mud Density - 9.6 ppg
Mud Density - 0.5 psi/ft

Depth - 10,000 feet

Figure 2.2 - Wellbore Schematic - Closed Container

Required:

1.   Determine the pressure in the gas bubble at point 1.

2.   Assuming that the 1 cubic foot of gas migrates to the surface of the closed container (point 2) with a constant volume of 1 cubic foot, determine the pressure at the

surface, the pressure at 2,000 feet, and the pressure at 10,000 feet.

Solution:

1.   The pressure of the gas, P_1, at point 1, which is the bottom of the hole, is determined by multiplying the gradient of the mud (psi/ft) by the depth of the well.

$$P_1 = \rho_m D \tag{2.5}$$

$$P_1 = 0.5\,(10000)$$

$$P_1 = 5000 \text{ psi}$$

2.   The pressure in the 1 cubic foot of gas at the surface (point 2) is determined using Equation 2.4:

$$P_1 V_1 = P_2 V_2$$

$$(5000)(1) = P_2(1)$$

$$P_{surface} = 5000 \text{ psi}$$

Determine the pressure at 2,000 feet:

$$P_{2000} = P_2 + \rho_m(2000)$$

$$P_{2000} = 5000 + 0.5(2000)$$

$$P_{2000} = 6000 \text{ psi}$$

Determine the pressure at the bottom of the hole.

$$P_{10000} = \rho_m(10000)$$

$$P_{10000} = 5000 + 0.5(10000)$$

$$P_{10000} = \textbf{10000 psi}$$

As illustrated in Example 2.2, the pressures in the well become excessive when the gas is not permitted to expand. The pressure at 2,000 feet would build to 6000 psi if the wellbore was a closed container. However, the wellbore is not a closed container and the pressure required to fracture the wellbore at 2,000 feet is 1520 psi. When the pressure at 2,000 feet exceeds 1520 psi, the container will rupture, resulting in an underground blowout.

The goal in circulating out a gas influx is to bring the gas to the surface, allowing the gas to expand to avoid rupturing the wellbore. At the same time, there is the need to maintain the total hydrostatic pressure at the bottom of the hole at the reservoir pressure in order to prevent additional influx of formation fluids. As will be seen, classical pressure control procedures routinely honor the second condition of maintaining the total hydrostatic pressure at the bottom of the hole equal to the reservoir pressure and ignore any consideration of the fracture pressure at the shoe.

The U - Tube Model

All classical displacement procedures are based on the U-Tube Model illustrated in Figure 2.3. It is important to understand this model and premise. Too often, field personnel attempt to apply classical well control procedures to non-classical problems. If the U-Tube Model does not accurately describe the system, classical pressure control procedures cannot be relied upon.

As illustrated in Figure 2.3, the left side of the U-Tube represents the drillpipe while the right side of the U-Tube represents the annulus. Therefore, the U-Tube Model describes a system where the bit is on bottom and it is possible to circulate from bottom. If it is not possible to circulate from bottom, classical well control concepts are meaningless and not applicable. This concept is discussed in detail in Chapter 4.

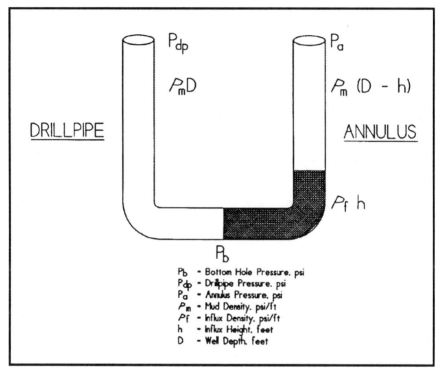

Figure 2.3 - The U-Tube Model

As further illustrated in Figure 2.3, an influx of formation fluids has entered the annulus (right side of the U-Tube). The well has been shut in, which means that the system has been closed. Under these shut-in conditions, there is static pressure on the drillpipe, which is denoted by P_{dp}, and static pressure on the annulus which is denoted by P_a. The formation fluid, P_f, has entered the annulus and occupies a volume defined by the area of the annulus and the height, h, of the influx.

An inspection of Figure 2.3 indicates that the drillpipe side of the U-Tube Model is more simple to analyze since the pressures are only influenced by mud of known density and pressure on the drillpipe that is easily measured. Under static conditions, the bottomhole pressure is easily determined utilizing Equation 2.6:

$$P_b = \rho_m D + P_{dp} \tag{2.6}$$

Where:

P_b = Bottomhole pressure, psi

ρ_m = Mud gradient, psi/ft

D = Well depth, feet

P_{dp} = Shut-in drillpipe pressure, psi

Equation 2.6 describes the shut-in bottomhole pressure in terms of the total hydrostatic on the drillpipe side of the U-Tube Model. The shut-in bottomhole pressure can also be described in terms of the total hydrostatic pressure on the annulus side of the U-Tube Model as illustrated by Equation 2.7:

$$P_b = \rho_f h + \rho_m (D - h) + P_a \tag{2.7}$$

Where:

P_b = Bottomhole pressure, psi

ρ_m = Mud gradient, psi/ft

D = Well depth, feet

P_a = Shut-in casing pressure, psi

ρ_f = Gradient of influx, psi/ft

h = Height of the influx, feet

Classic well control procedures, no matter what terminology is used, must keep the shut-in bottom hole pressure, P_b, constant to prevent additional influx of formation fluids while displacing the initial influx to the surface. Obviously, the equation for the drillpipe side (Equation 2.6) is the simpler and all of the variables are known; therefore, the drillpipe side is used to control the bottomhole pressure, P_b.

With the advent of pressure control technology, the necessity of spreading that technology presented an awesome task. Simplicity was in order and the classic Driller's Method for displacing the influx from the wellbore without permitting additional influx was developed.

DRILLER'S METHOD

The Driller's Method of displacement is simple and requires minimal calculations. The recommended procedure is as follows:

Step 1

On each tour, read and record the standpipe pressure at several rates in strokes per minute (spm), including the anticipated kill rate for each pump.

Step 2

After a kick is taken and prior to pumping, read and record the drillpipe and casing pressures. Determine the anticipated pump pressure at the kill rate using Equation 2.8:

$$P_c = P_{ks} + P_{dp} \tag{2.8}$$

Where:

P_c = Circulating pressure during displacement, psi
P_{ks} = Recorded pump pressure at the kill rate, psi
P_{dp} = Shut-in drillpipe pressure, psi

Important! If in doubt at any time during the entire procedure, shut in the well, read and record the shut-in drillpipe pressure and the shut-in casing pressure and proceed accordingly.

Step 3

Bring the pump to a kill speed, keeping the casing pressure constant at the shut-in casing pressure. This step should require less than five minutes.

Step 4

Once the pump is at a satisfactory kill speed, read and record the drillpipe pressure. Displace the influx, keeping the recorded drillpipe pressure constant.

Step 5

Once the influx has been displaced, record the casing pressure and compare with the original shut-in drillpipe pressure recorded in Step 1. It is important to note that, if the influx has been completely displaced, the casing pressure should be equal to the original shut-in drillpipe pressure.

Step 6

If the casing pressure is equal to the original shut-in drillpipe pressure recorded in Step 1, shut in the well by keeping the casing pressure constant while slowing the pumps. If the casing pressure is greater than the original shut-in drillpipe pressure, continue circulating for an additional circulation, keeping the drillpipe pressure constant and then shut in the well, keeping the casing pressure constant while slowing the pumps.

Step 7

Read, record, and compare the shut-in drillpipe and casing pressures. If the well has been properly displaced, the shut-in drillpipe pressure should be equal to the shut-in casing pressure.

Step 8

If the shut-in casing pressure is greater than the shut-in drillpipe pressure, repeat Steps 2 through 7.

Step 9

If the shut-in drillpipe pressure is equal to the shut-in casing pressure, determine the density of the kill-weight mud, ρ_1, using Equation 2.9 (Note that no "safety factor" is recommended or included):

$$\rho_1 = \frac{\rho_m D + P_{dp}}{0.052 D} \qquad (2.9)$$

Where:

ρ_1 = Density of the kill-weight mud, ppg

ρ_m = Gradient of the original mud, psi/ft
P_{dp} = Shut-in drillpipe pressure, psi
D = Well depth, feet

Step 10

Raise the mud weight in the suction pit to the density determined in Step 9.

Step 11

Determine the number of strokes to the bit by dividing the capacity of the drill string in barrels by the capacity of the pump in barrels per stroke according to Equation 2.10:

$$STB = \frac{C_{dp}l_{dp} + C_{hw}l_{hw} + C_{dc}l_{dc}}{C_p}$$ (2.10)

Where:

STB = Strokes to the bit, strokes
C_{dp} = Capacity of the drillpipe, bbl/ft
C_{hw} = Capacity of the heavy-weight drillpipe, bbl/ft
C_{dc} = Capacity of the drill collars, bbl/ft
l_{dp} = Length of the drillpipe, feet
l_{hw} = Length of the heavy-weight drillpipe, feet
l_{dc} = Length of the drill collars, feet
C_p = Pump capacity, bbl/stroke

Step 12

Bring the pump to speed, keeping the casing pressure constant.

Step 13

Displace the kill-weight mud to the bit, keeping the casing pressure constant.

Warning! Once the pump rate has been established, no further adjustments to the choke should be required. The casing pressure should remain constant at the initial shut-in drillpipe pressure. If the casing pressure begins to rise, the procedure should be terminated and the well shut in.

Step 14

After pumping the number of strokes required for the kill mud to reach the bit, read and record the drillpipe pressure.

Step 15

Displace the kill-weight mud to the surface, keeping the drillpipe pressure constant.

Step 16

With kill-weight mud to the surface, shut in the well by keeping the casing pressure constant while slowing the pumps.

Step 17

Read and record the shut-in drillpipe pressure and the shut-in casing pressure. Both pressures should be 0.

Step 18

Open the well and check for flow.

Step 19

If the well is flowing, repeat the procedure.

Step 20

If no flow is observed, raise the mud weight to include the desired trip margin and circulate until the desired mud weight is attained throughout the system.

The discussion of each step in detail follows:

Step 1

On each tour, read and record the standpipe pressure at several flow rates in strokes per minute (spm), including the anticipated kill rate for each pump.

Experience has shown that one of the most difficult aspects of any kill procedure is bringing the pump to speed without permitting an additional influx or fracturing the casing shoe. This problem is compounded by attempts to achieve a precise kill rate. There is nothing magic about the kill rate used to circulate out a kick. In the early days of pressure control, surface facilities were inadequate to bring an influx to the surface at a high pump speed. Therefore, one-half normal speed became the arbitrary rate of choice for circulating the influx to the surface. However, if only one rate such as the one-half speed is acceptable, problems can arise when the pump speed is slightly less or slightly more than the precise one-half speed. The reason for the potential problem is that the circulating pressure at rates other than the kill rate is unknown. Refer to further discussion after Step 4.

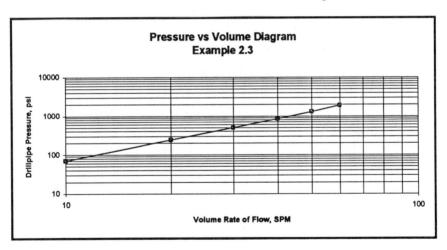

Figure 2.4

The best procedure is to record and graph several flow rates and corresponding pump pressures as illustrated in Figure 2.4. It is assumed in Examples 2.3 and 2.4 that the kill speed used is 30 strokes per minute. However, the actual pump speed used need not be exactly 30 strokes per minute. The drillpipe pressure corresponding to the actual pump speed being used could be verified readily using Figure 2.4.

Step 2

After a kick has been taken and prior to pumping, read and record the drillpipe and casing pressures. Determine the pump pressure at the kill speed.

Important! If in doubt at any time during the entire procedure, shut in the well, read and record the shut-in drillpipe pressure and the shut-in casing pressure and proceed accordingly.

It is not uncommon for the surface pressures to fluctuate slightly due to temperature, gas migration, or gauge problems. Therefore, for future reference it is important to record the surface pressures immediately prior to commencing pumping operations.

The second statement is extremely important to keep in mind. When in doubt, shut in the well! It seems that the prevailing impulse is to continue circulating regardless of the consequences. If the condition of the well has deteriorated since it was shut in, it deteriorated during the pumping phase. When in doubt, shut in the well, read the surface pressures, compare with the original pressures and evaluate the situation prior to further operations. If something is wrong with the displacement procedure being used, the situation is less likely to deteriorate while shut in and more likely to continue to deteriorate if pumping is continued. The well was under control when initially shut in.

Step 3

Bring the pump to a kill speed, keeping the casing pressure constant. This step should require less than five minutes.

As previously stated, bringing the pump to speed is one of the most difficult problems in any well control procedure. Experience has shown that the most practical approach is to keep the casing pressure constant at the shut-in casing pressure while bringing the pump to speed. The initial gas expansion is negligible over the allotted time of five minutes required to bring the pump to speed.

It is not important that the initial volume rate of flow be exact. Any rate within 10% of the kill rate is satisfactory. This procedure will establish the correct drillpipe pressure to be used to displace the kick. Figure 2.4 can be used to verify the drillpipe pressure being used.

Practically, the rate can be lowered or raised at any time during the displacement procedure. Simply read and record the circulating casing pressure and hold that casing pressure constant while adjusting the pumping rate and establishing a new drillpipe pressure. No more than one to two minutes can be allowed for changing the rate when the gas influx is near the surface because the expansion near the surface is quite rapid.

Step 4

Once the pump is at a satisfactory kill speed, read and record the drillpipe pressure. Displace the influx, keeping the recorded drillpipe pressure constant.

Actually, all steps must be considered together and are integral to each other. The correct drillpipe pressure used to circulate out the influx will be that drillpipe pressure established by Step 4. The pump rates and pressures established in Step 1 are to be used as a confirming reference only once the operation has commenced. Consideration of the U-Tube Model in Figure 2.3 clearly illustrates that, by holding the casing pressure constant at the shut-in casing pressure while bringing the pump to speed, the appropriate drillpipe pressure will be established for the selected rate.

All adjustments to the circulating operation must be performed considering the casing annulus pressure. In adjusting the pressure on the circulating system, the drillpipe pressure response must be considered secondarily because there is a significant lag time between any choke operation and the response on the drillpipe pressure gauge. This lag time is caused by the time required for the pressure transient to travel from the choke to the drillpipe pressure gauge. Pressure responses travel at the speed of sound in the medium. The speed of sound is 1,088 feet per second in air and about 4,800 feet per second in most water-based drilling muds. Therefore, in a 10,000-foot well, a pressure transient caused by opening or closing the choke would not be reflected on the standpipe pressure gauge until four seconds later. Utilizing only the drillpipe pressure and the choke usually results in large cyclical variations which cause additional influxes or unacceptable pressures at the casing shoe.

Step 5

Once the influx has been displaced, record the casing pressure and compare with the original shut-in drillpipe pressure recorded in Step 1. It is important to note that, if the influx has been completely displaced, the casing pressure should be equal to the original shut-in drillpipe pressure.

Consider the U-Tube Model presented in Figure 2.5 and compare with the U-Tube Model illustrated in Figure 2.3. If the influx has been properly and completely displaced, the conditions in the annulus side of Figure 2.5 are exactly the same as the conditions in the drillpipe side of Figure 2.3. If the frictional pressure losses in the annulus are negligible, the conditions in the annulus side of Figure 2.5 will be approximately the same as the drillpipe side of Figure 2.3. Therefore, once the influx is displaced, the circulating annulus pressure should be equal to the initial shut-in drillpipe pressure.

Step 6

If the casing pressure is equal to the original shut-in drillpipe pressure recorded in Step 1, shut in the well by keeping the casing pressure constant while slowing the pumps. If the casing pressure is greater than the original shut-in drillpipe pressure, continue circulating for an additional circulation, keeping the drillpipe pressure constant and then shut in the well, keeping the casing pressure constant while slowing the pumps.

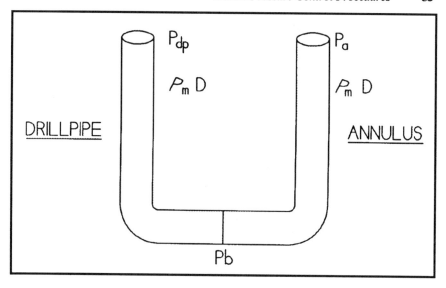

Figure 2.5 - Influx Displaced - The U-Tube Model

Step 7

Read, record, and compare the shut-in drillpipe and casing pressures. If the well has been properly displaced, the shut-in drillpipe pressure should be equal to the shut-in casing pressure.

Again consider Figure 2.5. Assuming that the influx has been completely displaced, conditions in both sides of the U-Tube Model are exactly the same. Therefore, the pressures at the surface on both the drillpipe and casing should be exactly the same.

Often, pressure is trapped in the system during the displacement procedure. If the drillpipe pressure and casing pressure are equal after displacing the influx but greater than the original shut-in drillpipe pressure or that drillpipe pressure recorded in Step 2, the difference between the two values is probably due to trapped pressure.

If the surface pressures recorded after displacement are equal but greater than the initial shut-in drillpipe pressure and formation influx is still present in the annulus, this discussion is not valid.

These conditions are discussed in the special problems in Chapter 4.

Step 8

If the shut-in casing pressure is greater than the shut-in drillpipe pressure, repeat Steps 2 through 7.

If, after displacing the initial influx, the shut-in casing pressure is greater than the shut-in drillpipe pressure, it is probable that an additional influx was permitted at some point during the displacement procedure.   Therefore, it will be necessary to displace that second influx.

Step 9

If the shut-in drillpipe pressure is equal to the shut-in casing pressure, determine the density of the kill-weight mud, ρ_1, using Equation 2.9 (Note that no "safety factor" is recommended or included):

$$\rho_1 = \frac{\rho_m D + P_{dp}}{0.052 D}$$

Safety factors are discussed in detail in Chapter 4.

Step 10

Raise the mud weight in the suction pit to the density determined in Step 9.

Step 11

Determine the number of strokes to the bit by dividing the capacity of the drill string in barrels by the capacity of the pump in barrels per stroke according to Equation 2.10:

$$STB = \frac{C_{dp} l_{dp} + C_{hw} l_{hw} + C_{dc} l_{dc}}{C_p}$$

Sections or different weights of drillpipe, drill collars or heavy weight drillpipe may be added or deleted from Equation 2.10 simply by adding to or subtracting from the numerator of Equation 2.10 the product of the capacity and the length of the section.

Step 12

Bring the pump to speed, keeping the casing pressure constant.

Step 13

Displace the kill-weight mud to the bit, keeping the casing pressure constant.

Warning! Once the pump rate has been established, no further adjustments to the choke should be required. The casing pressure should remain constant at the initial shut-in drillpipe pressure. If the casing pressure begins to rise, the procedure should be terminated and the well shut in.

It is vital to understand Step 13. Again, consider the U-Tube Model in Figure 2.5. While the kill-weight mud is being displaced to the bit on the drillpipe side, under dynamic conditions no changes are occurring in any of the conditions on the annulus side. Therefore, once the pump rate has been established, the casing pressure should not change and it should not be necessary to adjust the choke to maintain the constant drillpipe pressure. If the casing pressure does begin to increase, with everything else being constant, in all probability there is some gas in the annulus. If there is gas in the annulus, this procedure must be terminated. Since the density of the mud at the surface has been increased to the kill, the proper procedure under these conditions would be the Wait and Weight Method. Therefore, the Wait and Weight Method would be used to circulate the gas in the annulus to the surface and control the well.

Step 14

After pumping the number of strokes required for the kill mud to reach the bit, read and record the drillpipe pressure.

Step 15

Displace the kill-weight mud to the surface, keeping the drillpipe pressure constant.

Referring to Figure 2.5, once kill-weight mud has reached the bit and the displacement of the annulus begins, conditions on the drillpipe side of the U-Tube Model are constant and do not change. Therefore, the kill-weight mud can be displaced to the surface by keeping the drillpipe pressure constant. Some change in casing pressure and adjustment in the choke size can be expected during this phase. If the procedure has been executed properly, the choke size will be increased to maintain the constant drillpipe pressure and the casing pressure will decline to 0 when the kill-weight mud reaches the surface.

Step 16

With kill-weight mud to the surface, shut in the well by keeping the casing pressure constant while slowing the pumps.

Step 17

Read and record the shut-in drillpipe pressure and the shut-in casing pressure. Both pressures should be 0.

Step 18

Open the well and check for flow.

Step 19

If the well is flowing, repeat the procedure.

Step 20

If no flow is observed, raise the mud weight to include the desired trip margin and circulate until the desired mud weight is attained throughout the system.

The Driller's Method is illustrated in Example 2.3:

Example 2.3
   Given:

Wellbore schematic		=	Figure 2.6
Well depth,	D	=	10,000 feet
Hole size,	D_h	=	7 7/8 inches
Drillpipe size,	D_p	=	4 ½ inches
8 5/8-inch surface casing		=	2,000 feet
Casing internal diameter,	D_{ci}	=	8.017 inches
Fracture gradient,	F_g	=	0.76 psi/ft
Mud weight,	ρ	=	9.6 ppg
Mud gradient,	ρ_m	=	0.50 psi/ft

A kick is taken with the drill string on bottom and

Shut-in drillpipe pressure,	P_{dp}	=	200 psi
Shut-in annulus pressure,	P_a	=	300 psi
Pit level increase		=	10 barrels
Normal circulation rate		=	6 bpm at 60 spm
Kill rate		=	3 bpm at 30 spm
Circulating pressure at kill rate, P_{ks}		=	500 psi

Pump capacity, C_p =   0.1 bbl/stk

Capacity of the:

Drillpipe, C_{dpi} =   0.0142 bbl/ft

Drillpipe casing annulus, C_{dpca} =   0.0428 bbl/ft

Drillpipe hole annulus, C_{dpha} =   0.0406 bbl/ft

Note:

For simplicity in calculation and illustration, no drill collars are assumed. The inclusion of drill collars adds only an intermediate calculation.

Required:
Describe the kill procedure using the Driller's Method.

Solution:
  1.   Establish pressure versus volume diagram (Figure 2.4).

  2.   Record the shut-in drillpipe pressure and shut-in casing pressure.

$P_{dp} = 200$ psi

$P_a = 300$ psi

  3.   Establish the pumping pressure at the kill rate of 30 spm using Equation 2.8:

$$P_c = P_{ks} + P_{dp}$$

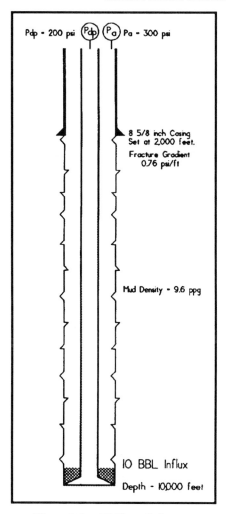

Figure 2.6 - Wellbore Schematic

$$P_c = 500 + 200$$

$$P_c = 700 \text{ psi}$$

4.   Bring the pump to 30 spm, maintaining 300 psi on the casing annulus.

5.   Read and record the drillpipe pressure equal to 700 psi. Confirm the drillpipe pressure using Figure 2.4.

6.   Displace the annulus and all gas which has entered the wellbore, keeping the drillpipe pressure constant at 700 psi.

7.   Read and record the drillpipe pressure equal to 700 psi and casing pressure equal to 200 psi.

8.   Shut in the well, keeping the casing pressure constant at 200 psi. Allow the well to stabilize.

9.   Determine that all gas is out of the mud.

$$P_a = P_{dp} = \textbf{200 psi.}$$

10.   Determine the kill weight, ρ_1, using Equation 2.9:

$$\rho_1 = \frac{\rho_m D + P_{dp}}{0.052 D}$$

$$\rho_1 = \frac{0.5(10000) + 200}{0.052(10000)}$$

$$\rho_1 = \textbf{10 ppg}$$

11.   Determine the strokes to the bit using Equation 2.10:

$$STB = \frac{C_{dp} l_{dp} + C_{hw} l_{hw} + C_{dc} l_{dc}}{C_p}$$

$$STB = \frac{(0.0142)(10000)}{0.1}$$

STB = **1,420 strokes**

12.   Raise the mud weight at the surface to 10 ppg.

13.   Bring the pump to 30 spm, keeping the casing pressure constant at 200 psi.

14.   Displace the 10-ppg mud to the bit with 1,420 strokes, keeping the annulus pressure, P_a, constant at 200 psi. The choke size must not change.

15.   At 1,420 strokes observe and record the circulating pressure on the drillpipe. Assume that the observed pressure on the drillpipe is 513 psi.

16.   Circulate the 10-ppg mud to the surface, keeping the drillpipe pressure constant at 513 psi.

17.   Shut in and check $P_a = P_{dp} = 0$ psi. The well is dead.

18.   Circulate and raise the mud weight to some acceptable trip margin, generally between 150 to 500 psi above the formation pressure, P_b, or 10.3 to 11.0 ppg.

19.   Continue drilling ahead.

WAIT AND WEIGHT METHOD

The alternative classical method is commonly known as the Wait and Weight Method. As the name implies, the well is shut in while the mud density is increased to the kill weight as determined by Equation 2.9. Therefore, the primary difference is operational in that the kill-weight mud, ρ_1, is pumped while the gas is being displaced. The result is that the well is killed in one circulation with the Wait and Weight Method whereas, with the Driller's Method, two circulations are required. In the early days of pressure control, the time required to increase the density of the mud in the surface system to the kill weight was significant. During that time, it was not uncommon for the gas to migrate or for the drillpipe to become stuck. However, modern mud mixing systems have eliminated the time factor from most operations in that most systems can raise the density of the surface system as fast as the mud is pumped. There are other important comparisons which will be presented after the Wait and Weight Method is presented, illustrated and discussed.

The displacement procedure for the Wait and Weight Method is as follows, subsequently, each step will be discussed in detail:

Step 1

On each tour read and record the standpipe pressure at several rates in strokes per minute (spm), including the anticipated kill rate for each pump.

Step 2

Prior to pumping, read and record the drillpipe and casing pressures. Determine the anticipated pump pressure at the kill rate using Equation 2.8:

$$P_c = P_{ks} + P_{dp}$$

Step 3

Determine the density of the kill-weight mud, ρ_1, using Equation 2.9 (note that no "safety factor" is recommended or included):

$$\rho_1 = \frac{\rho_m D + P_{dp}}{0.052 D}$$

Step 4

Determine the number of strokes to the bit by dividing the capacity of the drill string in barrels by the capacity of the pump in barrels per stroke according to Equation 2.10:

$$STB = \frac{C_{dp}l_{dp} + C_{hw}l_{hw} + C_{dc}l_{dc}}{C_p}$$

Step 5

Determine the new circulating pressure, P_{cn}, at the kill rate with the kill-weight mud at the bit utilizing Equation 2.11:

$$P_{cn} = P_{dp} - 0.052(\rho_1 - \rho)D + \left(\frac{\rho_1}{\rho}\right)P_{ks} \qquad (2.11)$$

Where:

ρ_1 = Density of the kill-weight mud, ppg

ρ = Density of the original mud, ppg

P_{ks} = Original circulating pressure at kill rate, psi

P_{dp} = Shut-in drillpipe pressure, psi

D = Well depth, feet

Step 6

For a complex drill string configuration, determine and graph the pumping schedule for reducing the initial circulating pressure, P_c, determined in Step 2 to the final circulating pressure, P_{cn}, determined in Step 5. Using Equations 2.12 and 2.13, calculate Table 1 and create the corresponding graph.

Note: A "section" of drill string is the length where all the diameters remain the same. A new section would start anytime the hole size or pipe diameter changed. As long as the diameters remain the same, it is one section. Therefore, each section has one annular capacity. The calculations begin from the surface.

For example, if the hole size does not change and the string consists of two weights of drillpipe, heavy-weight drillpipe and drill collars, four calculations would be required.

The table would look like the following:

Strokes	Pressure
0	700
STKS1	P_1
STKS2	P_2
STKS3	P_3
...	...
STB	P_{cn}

$$STKS\,1 = \frac{C_{ds1}l_{ds1}}{C_p} \qquad (2.12a)$$

$$STKS\,2 = \frac{C_{ds1}l_{ds1} + C_{ds2}l_{ds2}}{C_p} \qquad (2.12b)$$

$$STKS\,3 = \frac{C_{ds1}l_{ds1} + C_{ds2}l_{ds2} + C_{ds3}l_{ds3}}{C_p} \qquad (2.12c)$$

$$STB = \frac{C_{ds1}l_{ds1} + C_{ds2}l_{ds2} + C_{ds3}l_{ds3} + ... + C_{dc}l_{dc}}{C_p} \qquad (2.12d)$$

$$P_1 = P_c - 0.052(\rho_1 - \rho)(l_{ds1}) + \left(\frac{\rho_1 P_{ks}}{\rho} - P_{ks}\right)\left(\frac{STKS\,1}{STB}\right) \qquad (2.13a)$$

$$P_2 = P_c - 0.052(\rho_1 - \rho)(l_{ds1} + l_{ds2}) +$$
$$\left(\frac{\rho_1 P_{ks}}{\rho} - P_{ks}\right)\left(\frac{STKS\,2}{STB}\right) \qquad (2.13b)$$

$$P_3 = P_c - 0.052(\rho_1 - \rho)(l_{ds1} + l_{ds2} + l_{ds3}) +$$
$$\left(\frac{\rho_1 P_{ks}}{\rho} - P_{ks}\right)\left(\frac{STKS\,3}{STB}\right) \qquad (2.13c)$$

$$P_{cn} = P_{dp} - 0.052(\rho_1 - \rho)*$$
$$(l_{ds1} + l_{ds2} + l_{ds3} + ... + l_{dc}) + \left(\frac{\rho_1 P_{ks}}{\rho} - P_{ks}\right) \qquad (2.13d)$$

Where:

    STKS1 = Strokes to end of section 1 of drill string
    STKS2 = Strokes to end of section 2 of drill string
    STKS3 = Strokes to end of section 3 of drill string
    STB   = Strokes to the bit as determined in Step 4

Where:

ρ_1	= Density of kill-weight mud, ppg
ρ	= Density of original mud, ppg
$l_{ds1,2,3}$	= Length of section of drill string, feet
$C_{ds1,2,3}$	= Capacity of section of drill string, bbl/ft
$P_{1,2,3}$	= Circulating pressure with kill-weight mud to the end of section 1,2,3, psi
P_{dp}	= Shut-in drillpipe pressure, psi
P_{ks}	= Circulating pressure at kill speed determined in Step 1, psi
C_p	= Pump capacity, bbl/stroke
P_{cn}	= New circulating pressure, psi
P_c	= Initial displacement pressure determined in Step 2 using Equation 2.8, psi

For a drill string composed of only one weight of drillpipe and one string of heavy-weight drillpipe or drill collars, the pumping schedule can be determined using Equation 2.14:

$$\frac{STKS}{25\,psi} = \frac{25(STB)}{P_c - P_{cn}}$$

(2.14)

Step 7

Raise the density of the mud in the suction pit to the kill weight determined in Step 3.

Step 8

Bring the pump to a kill speed, keeping the casing pressure constant at the shut-in casing pressure. This step should require less than five minutes.

Step 9

Once the pump is at a satisfactory kill speed, read and record the drillpipe pressure. Adjust the pumping schedule accordingly. Verify the drillpipe pressure using the diagram established in Step 1. Displace the kill-weight mud to the bit pursuant to the pumping schedule established in Step 6 as revised in this step.

Step 10

Displace the kill-weight mud to the surface, keeping the drillpipe pressure constant.

Step 11

Shut in the well, keeping the casing pressure constant and observe that the drillpipe pressure and the casing pressure are 0 and the well is dead.

Step 12

> If the surface pressures are not 0 and the well is not dead, continue to circulate, keeping the drillpipe pressure constant.

Step 13

> Once the well is dead, raise the mud weight in the suction pit to provide the desired trip margin.

Step 14

> Drill ahead.

> Discussion of each step in detail follows:

Step 1

> On each tour, read and record the standpipe pressure at several rates in strokes per minute (spm), including the anticipated kill rate for each pump.

> This is the same discussion as presented after Step 1 of the Driller's Method. Experience has shown that one of the most difficult aspects of any kill procedure is bringing the pump to speed without permitting an additional influx or fracturing the casing shoe. This problem is compounded by attempts to achieve a precise kill rate. There is nothing magic about the kill rate used to circulate out a kick. In the early days of pressure control, surface facilities were inadequate to bring an influx to the surface at a high pump speed. Therefore, one-half normal speed became the arbitrary rate of choice for circulating the influx to the surface. However, if only one rate such as the one-half speed is acceptable, problems can arise when the pump speed is slightly less or slightly more than the precise one-half speed. The reason for the potential problem is that the circulating pressure at rates other than the kill rate is unknown. Refer to further discussion after Step 4.

> The best procedure is to record and graph several flow rates and corresponding pump pressures as illustrated in Figure 2.4. It is assumed in Examples 2.3 and 2.4 that the kill speed used is 30

strokes per minute. However, the actual pump speed used need not be exactly 30 strokes per minute. The drillpipe pressure corresponding to the actual pump speed being used could be verified readily using Figure 2.4.

Step 2

Prior to pumping, read and record the drillpipe and casing pressures. Determine the anticipated pump pressure at the kill rate using Equation 2.8:

$$P_c = P_{ks} + P_{dp}$$

This is the same discussion as presented after Step 2 of the Driller's Method. It is not uncommon for the surface pressures to fluctuate slightly due to temperature, migration or gauge problems. Therefore, for future reference it is important to record the surface pressures immediately prior to commencing pumping operations.

When in doubt, shut in the well! It seems that the prevailing impulse is to continue circulating regardless of the consequences. If the condition of the well has deteriorated since it was shut in, it deteriorated during the pumping phase. When in doubt, shut in the well, read the surface pressures, compare with the original pressures and evaluate the situation prior to further operations. If something is wrong with the displacement procedure being used, the situation is less likely to deteriorate while shut in and more likely to continue to deteriorate if pumping is continued. The well was under control when initially shut in.

Step 3

Determine the density of the kill-weight mud, ρ_1, using Equation 2.9 (note that no "safety factor" is recommended or included):

$$\rho_1 = \frac{\rho_m D + P_{dp}}{0.052 D}$$

Safety factors are discussed in Chapter 4.

Step 4

Determine the number of strokes to the bit by dividing the capacity of the drill string in barrels by the capacity of the pump in barrels per stroke according to Equation 2.10:

$$STB = \frac{C_{dp} l_{dp} + C_{hw} l_{hw} + C_{dc} l_{dc}}{C_p}$$

Sections of different weights of drillpipe, drill collars, or heavy-weight drillpipe may be added or deleted from Equation 2.10 simply by adding to or subtracting from the numerator of Equation 2.10 the product of the capacity and the length of the section.

Step 5

Determine the new circulating pressure, P_{cn}, at the kill rate with the kill-weight mud utilizing Equation 2.11:

$$P_{cn} = P_{dp} - 0.052(\rho_1 - \rho)D + \left(\frac{\rho_1}{\rho}\right) P_{ks}$$

The new circulating pressure with the kill-weight mud will be slightly greater than the recorded circulating pressure at the kill speed since the frictional pressure losses are a function of the density of the mud. In Equation 2.11 the frictional pressure loss is considered a direct function of the density. In reality, the frictional pressure loss is a function of the density to the 0.8 power. However, the difference is insignificant.

Step 6

For a complex drill string configuration, determine and graph the pumping schedule for reducing the initial circulating pressure, P_c, determined in Step 2 to the final circulating pressure, P_{cn}, determined in Step 5. Using Equations 2.12 and 2.13, calculate Table 1 and create the corresponding graph.

Note: A "section" of drill string is the length where all the diameters remain the same. A new section would start anytime the hole size or pipe diameter changed. As long as the diameters remain the same, it is one section. Therefore, each section has one annular capacity. The calculations begin from the surface. For example, if the hole size does not change and the string consists of two weights of drillpipe, heavy-weight drillpipe and drill collars, four calculations would be required.

Determining this pump schedule is a most critical phase. Use of these equations is illustrated in Examples 2.3 and 2.4. Basically, the circulating drillpipe pressure is reduced systematically to offset the increase in hydrostatic introduced by the kill-weight mud and ultimately to keep the bottomhole pressure constant.

The systematic reduction in drillpipe pressure must be attained by reducing the casing pressure by the scheduled amount and waiting four to five seconds for the pressure transient to reach the drillpipe pressure gauge. Efforts to control the drillpipe pressure directly by manipulating the choke are usually unsuccessful due to the time lag. The key to success is to observe several gauges at the same time. The sequence is usually to observe the choke position, the casing pressure and drillpipe pressure. Then concentrate on the choke position indicator while slightly opening the choke. Next, check the choke pressure gauge for the reduction in choke pressure. Continue that sequence until the designated amount of pressure has been bled from the annulus pressure gauge. Finally, wait 10 seconds and read the result on the drillpipe pressure gauge. Repeat the process until the drillpipe pressure has been adjusted appropriately.

Step 7

Raise the density of the mud in the suction pit to the kill weight determined in Step 3.

Step 8

Bring the pump to a kill speed, keeping the casing pressure constant at the shut-in casing pressure. This step should require less than five minutes.

Step 9

Once the pump is at a satisfactory kill speed, read and record the drillpipe pressure. Adjust the pumping schedule accordingly. Verify the drillpipe pressure using the diagram established in Step 1. Displace the kill-weight mud to the bit pursuant to the pumping schedule established in Step 6 as revised in this step.

As discussed in the Driller's Method, the actual kill speed used is not critical. Once the actual kill speed is established at a constant casing pressure equal to the shut-in casing pressure, the drillpipe pressure read is correct. The pumping schedule must be adjusted to reflect a pump speed different from the pump speed used to construct the table and graph. The adjustment of the table is accomplished by reducing arithmetically the initial drillpipe pressure by the shut-in drillpipe pressure and re-making the appropriate calculations. The graph is more easily adjusted. The circulating drillpipe pressure marks the beginning point. Using that point, a line is drawn which is parallel to the line drawn in Step 6. The new line becomes the correct pumping schedule. The graph of pump pressure versus volume constructed in Step 1 is used to confirm the calculations.

If doubt arises during the pumping procedure, the well should be shut in by keeping the casing pressure constant while slowing the pump. The shut-in drillpipe pressure, shut-in casing pressure, and volume pumped should be used to evaluate the situation. The pumping procedure can be continued by bringing the pump to speed keeping the casing pressure constant, reading the drillpipe pressure, plotting the point on the pumping schedule graph, and establishing a new line parallel to the original. These points are clarified in Examples 2.3 and 2.4.

Keeping the casing pressure constant in order to establish the pump speed and correct circulating drillpipe pressure is an acceptable procedure provided that the time period is short and the influx is not near the surface. The time period should never be more than five minutes. If the influx is near the surface, the casing pressures will be changing very rapidly. In that case, the time period should be one to two minutes.

Step 10

Displace the kill-weight mud to the surface, keeping the drillpipe pressure constant.

Step 11

Shut in the well, keeping the casing pressure constant and observe that the drillpipe pressure and the casing pressure are 0 and the well is dead.

Step 12

If the surface pressures are not 0 and the well is not dead, continue to circulate, keeping the drillpipe pressure constant.

Step 13

Once the well is dead, raise the mud weight in the suction pit to provide the desired trip margin.

Step 14

Drill ahead.

The Wait and Weight Method is illustrated in Example 2.4:

Example 2.4
Given:

Wellbore schematic   =   Figure 2.6

Well depth,	D	=	10,000 feet
Hole size,	D_h	=	7 7/8 inches
Drillpipe size,	D_p	=	4 ½ inches
8 5/8-inch surface casing		=	2,000 feet
Casing internal diameter,	D_{ci}	=	8.017 inches
Fracture gradient,	F_g	=	0.76 psi/ft
Mud weight,	ρ	=	9.6 ppg
Mud gradient,	ρ_m	=	0.50 psi/ft

A kick is taken with the drill string on bottom and

Shut-in drillpipe pressure,	P_{dp}	=	200 psi
Shut-in annulus pressure,	P_a	=	300 psi
Pit level increase		=	10 barrels
Normal circulation rate		=	6 bpm at 60 spm
Kill rate		=	3 bpm at 30 spm
Circulating pressure at kill rate,	P_{ks}	=	500 psi
Pump capacity,	C_p	=	0.1 bbl/stk

Capacity of the:

Drillpipe,	C_{dpi}	=	0.0142 bbl/ft
Drillpipe casing annulus,	C_{dpca}	=	0.0428 bbl/ft
Drillpipe hole annulus,	C_{dpha}	=	0.0406 bbl/ft

Note: For simplicity in calculation and illustration, no drill collars are assumed. The inclusion of drill collars adds only an intermediate calculation.

Required:

Describe the kill procedure using the Wait and Weight Method.

Solution:

1. Establish pressure versus volume diagram using Figure 2.4.

2. Record the shut-in drillpipe pressure and shut-in casing pressure:

$$P_{dp} = 200 \text{ psi}$$

$$P_a = 300 \text{ psi}$$

3. Establish the pumping pressure at the kill rate of 30 spm using Equation 2.8:

$$P_c = P_{ks} + P_{dp}$$

$$P_c = 500 + 200$$

$$P_c = 700 \text{ psi}$$

4. Determine the kill weight, ρ_1, using Equation 2.9:

$$\rho_1 = \frac{\rho_m D + P_{dp}}{0.052 D}$$

$$\rho_1 = \frac{0.5(10000) + 200}{0.052(10000)}$$

$$\rho_1 = 10 \text{ ppg}$$

5. Determine the strokes to the bit using Equation 2.10:

$$STB = \frac{C_{dp} l_{dp} + C_{hw} l_{hw} + C_{dc} l_{dc}}{C_p}$$

$$STB = \frac{(0.0142)(10000)}{0.1}$$

$$STB = 1{,}420 \text{ strokes}$$

6. Determine the new circulating pressure, P_{cn}, at the kill rate with the kill-weight mud utilizing Equation 2.11:

$$P_{cn} = P_{dp} - 0.052(\rho_1 - \rho)D + \left(\frac{\rho_1}{\rho}\right)P_{ks}$$

$$P_{cn} = 200 - 0.052(10 - 9.6)(10000) + \left(\frac{10}{9.6}\right)500$$

$$P_{cn} = 513 \text{ psi}$$

7.   Determine the pumping schedule for a simple drill string pursuant to Equation 2.14:

$$\frac{STKS}{25\,psi} = \frac{25(STB)}{P_c - P_{cn}}$$

$$\frac{STKS}{25\,psi} = \frac{25(1420)}{700 - 513}$$

$$\frac{STKS}{25\,psi} = 190 \text{ strokes}$$

Strokes	Pressure
0	700
190	675
380	650
570	625
760	600
950	575
1140	550
1330	525
1420	513
1600	513

8.   Construct Figure 2.7 — graph of pump schedule.

9.   Bring pump to 30 spm, keeping the casing pressure constant at 300 psi.

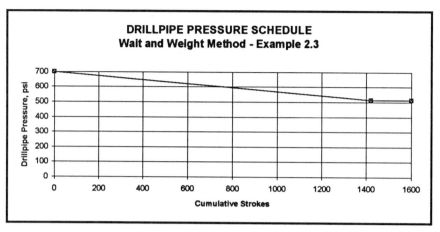

Figure 2.7

10. Displace the kill-weight mud to the bit (1,420 strokes) according to the pump schedule developed in Steps 7 and 8.

   After 190 strokes, reduce the casing pressure observed at that moment by 25 psi. After 10 seconds, observe that the drillpipe pressure has dropped to 675 psi. After 380 strokes, reduce the casing pressure observed at that moment by 25 psi. After 10 seconds, observe that the drillpipe pressure has dropped to 650 psi. Continue in this manner until the kill-weight mud is at the bit and the drillpipe pressure is 513 psi.

11. With the kill-weight mud to the bit after 1,420 strokes, read and record the drillpipe pressure equal to 513 psi.

12. Displace the kill-weight mud to the surface, keeping the drillpipe pressure constant at 513 psi.

13. Shut in the well, keeping the casing pressure constant and observe that the drillpipe and casing pressure are 0.

14. Check for flow.

15   Once the well is confirmed dead, raise the mud weight to provide the desired trip margin and drill ahead.

Obviously, the most potentially confusing aspect of the Wait and Weight Method is the development and application of the pumping schedule used to circulate the kill-weight mud properly to the bit while maintaining a constant bottomhole pressure. The development and application of the pump schedule is further illustrated in Example 2.5 to provide additional clarity:

Example 2.5

Given:

Example 2.4

Required:

1.   Assume that the kill speed established in Step 9 of Example 2.4 was actually 20 spm instead of the anticipated 30 spm. Determine the effect on the pump schedule and demonstrate the application of Figure 2.7.

2.   Assume that the drill string is complex and composed of 4,000 feet of 5-inch 19.5 #/ft, 4,000 feet of 4 ½-inch 16.6 #/ft, 1,000 feet of 4 ½-inch heavy weight drillpipe, and 1,000 feet of 6-inch-by-2-inch drill collars. Illustrate the effect of the complex string configuration on Figure 2.7. Compare the pump schedule developed for the complex string with that obtained with the straight line simplification.

Solution:

1.   Pursuant to Figure 2.4, the surface pressure at a kill speed of 20 spm would be 240 psi. The initial displacement surface pressure would be given by Equation 2.8 as follows:

$$P_c = P_{ks} + P_{dp}$$

$$P_c = 240 + 200$$

$$P_c = \textbf{440 psi}$$

Therefore, simply locate 440 psi on the Y axis of Figure 2.7 and draw a line parallel to that originally drawn. The new line is the revised pumping schedule. This concept is illustrated as Figure 2.8.

As an alternative, merely subtract 260 psi (700 - 440) from the values listed in the table in Step 7.

Strokes	Pressure
0	440
190	415
380	390
570	365
760	340
950	315
1140	290
1330	265
1420	253
1600	253

2.   Equations 2.10, 2.12 and 2.13 are used to graph the new pump schedule presented as Figure 2.9:

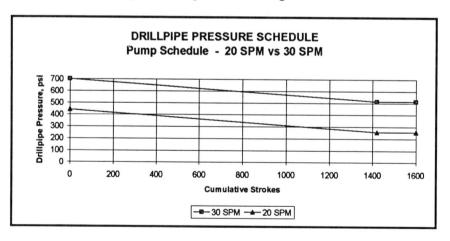

Figure 2.8

$$STB = \frac{C_{dp}l_{dp} + C_{hw}l_{hw} + C_{dc}l_{dc}}{C_p}$$

$$STB = \{(0.01776)(4000) +$$
$$(0.01422)(4000) + (0.00743)(1000) +$$
$$(0.00389)(1000)\} \div 0.10$$

$$STB = \textbf{1,392 strokes}$$

The graph is determined from Equations 2.12 and 2.13:

$$STKS\,1 = \frac{C_{dsl}l_{dsl}}{C_p}$$

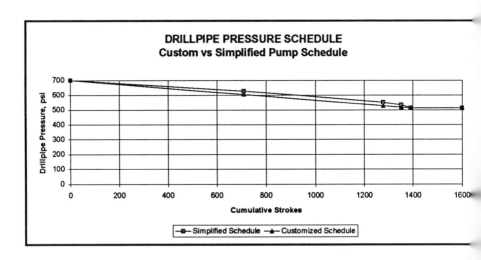

Figure 2.9

$$STKS\ 1 = \frac{(0.01776)(4000)}{0.1}$$

$STKS\ 1 =$ **710 strokes**

$$STKS\ 2 = \frac{C_{ds1}l_{ds1} + C_{ds2}l_{ds2}}{C_p}$$

$$STKS\ 2 = \frac{(0.01776)(4000) + (0.01422)(4000)}{0.1}$$

$STKS\ 2 =$ **1,279 strokes**

$$STKS\ 3 = \frac{C_{ds1}l_{ds1} + C_{ds2}l_{ds2} + C_{ds3}l_{ds3}}{C_p}$$

$$STKS3 = \{(0.01776)(4000) + (0.01422)(4000) + (0.00743)(1000)\} \div 0.10$$

$STKS\ 3 =$ **1,353 strokes**

Similarly, $STKS\ 4 =$ **1,392 strokes**

The circulating pressure at the surface at the end of each section of drill string is given by Equation 2.13:

$$P_1 = P_c - 0.052(\rho_1 - \rho)(l_{ds1}) + (\frac{\rho_1 P_{ks}}{\rho} - P_{ks})\left(\frac{STKS\ 1}{STB}\right)$$

$$P_1 = 700 - 0.052(10 - 9.6)(4000) +$$
$$\left(\frac{(10)(500)}{(9.6)} - 500 \right)\left(\frac{710}{1392} \right)$$

$$P_1 = 627 \text{ psi}$$

$$P_2 = P_c - 0.052(\rho_1 - \rho)(l_{ds1} + l_{ds2}) +$$
$$(\frac{\rho_1 P_{ks}}{\rho} - P_{ks})\left(\frac{STKS\ 2}{STB} \right)$$

$$P_2 = 700 - 0.052(10 - 9.6)(4000 + 4000) +$$
$$\left(\frac{(10)(500)}{(9.6)} - 500 \right)\left(\frac{1279}{1392} \right)$$

$$P_2 = 552 \text{ psi}$$

$$P_3 = P_c - 0.052(\rho_1 - \rho)(l_{ds1} + l_{ds2} + l_{ds3}) +$$
$$(\frac{\rho_1 P_{ks}}{\rho} - P_{ks})\left(\frac{STKS\ 3}{STB} \right)$$

$$P_3 = 700 - 0.052(10 - 9.6)(4000 + 4000 + 1000) +$$
$$\left(\frac{(10)(500)}{(9.6)} - 500 \right)\left(\frac{1353}{1392} \right)$$

$$P_3 = 534 \text{ psi}$$

Adding the the final section of drill collars to the above equation, $P_4 = 513$ psi

As illustrated in Figure 2.8, changing the pump speed at which the kick is displaced merely moves the pump schedule to a parallel position on the graph. As illustrated in Figure 2.9, the complex pump schedule is slightly more difficult to construct. The simplified straight line pump schedule approach will underbalance the well during the period that the kill-weight mud is being displaced to the bit. In this example the underbalance is only as much as 25 psi. In reality, in most cases the annular frictional pressure losses, which are considered negligible in classical pressure control analysis, would more than compensate and an additional influx would not occur. However, that may not be the case in any specific instance and an additional influx could occur. In most instances, the simplified pump schedule would suffice. In significantly complex drill strings, this comparison should be made.

SUMMARY

The Driller's Method was the first and most popular displacement procedure. The crew proceeded immediately to displace the influx. The required calculations were not difficult. The calculations were made, the kill-weight mud was easily displaced and the drilling operation was resumed. One disadvantage of the Driller's Method is that at least two circulations are required to control the well.

The Wait and Weight Method is slightly more complicated but offers some distinct advantages. First, the well is killed in half the time. Modern mud-mixing facilities permit barite to be mixed at rates up to 600 sacks per hour with dual mixing systems; therefore, time required to weight up the suction pit is minimized and kill rate is not penalized. The Wait and Weight Method results in kill mud in the well sooner and that is always an advantage. In addition, as discussed and illustrated in Chapter 4, the annulus pressures are lower when the Wait and Weight Method is used. The primary disadvantage is the potential for errors and problems while displacing the kill-weight mud to the bit. With the Driller's Method, the procedure can be stopped and started easily. Stopping and starting when using the Wait and Weight Method is not as easy, especially during the period that the kill-weight mud is being displaced to the bit. It is not uncommon that good drilling men get confused during displacement using the Wait and Weight Method.

In view of all considerations, the Wait and Weight Method is the preferred technique.

CHAPTER THREE
PRESSURE CONTROL PROCEDURES
WHILE TRIPPING

22 June

Trip out of the hole. Well began flowing. Trip in the hole with the bottomhole assembly. Gained 60 bbls. Circulated the hole. Pressure continued to increase. Shut in well with 3000 psi. Drillpipe started coming through the rotary table. Opened choke, closed pipe rams above tool joint, and closed choke. Total gain was 140 barrels. Attempt to close safety valve without success. Open choke, close safety valve, close choke. Had 200 barrel total gain. Rigged up snubbing unit.

14 July

Began snubbing in the hole.

15 July

Continued snubbing to 4,122 feet. Circulated 18.5-ppg mud. Surface pressure is 3700 psi.

16 July

Snubbed to 8,870 feet. Began to circulate 18.5-ppg mud. Surface pressure increased to 5300 psi on casing.

Mud in is 540 bbls versus mud out of 780 bbls. Increased pump rate from 3 bpm to 6 bpm. Drillpipe pressure increased to 6700 psi. Hammer union on rig floor washed out. Unable to close safety valves on rig floor due to excessive pressure and flow. Hydrogen sulfide monitors sounded. Ignited rig with flare gun.

The drilling report above is a good illustration of a disaster resulting from complications during tripping operations at a well that was under control only a few hours prior to the trip. It must be reasoned that any well that is tripped is under control when the bit leaves bottom. Therefore, operations subsequent to the beginning of the trip precipitate the well control problem. As is often the case, due to operations subsequent to the time that a well control problem was detected, the condition of the well deteriorated and the well control problem became progressively more complicated.

The reason for this familiar unfortunate chain of events is that the industry has been and still is inconsistent under these circumstances. A recent survey of the major well control schools revealed substantial inconsistencies under the same given circumstances.

Actually, classical pressure control procedures apply only to problems which occur during drilling operations. Unfortunately, there is no widely accepted procedure to be followed when a kick occurs during a trip. Further, procedures and instructions which apply to problems during drilling are routinely posted on the rig floor. However, just as routinely, there are no posted instructions which apply to procedures to be followed if the kick occurs during a trip. A federal court in Pecos, Texas, found a major oil company grossly negligent because procedures for problems occurring while drilling were posted and procedures for problems occurring while tripping were not. One purpose of this chapter is to suggest that classical trip procedures be adopted, taught and posted.

CAUSES OF KICKS WHILE TRIPPING

Any well control problem which occurs during a trip is generally the result of a failure on the part of the rig crew to keep the hole full or the failure of the crew to recognize that the hole is not filling properly. The problem of keeping the hole full of fluid has been emphasized for many years. Pressure control problems and blowouts associated with trips continue to be a major occurrence. A lack of training and understanding contributes to these circumstances. Classical pressure control procedures apply to drilling operations, not to tripping operations. All of the modeling and technology used in pressure control was developed based on a drilling model as opposed to a tripping model. Therefore, the technology that applies to pressure control problems which occur during drilling operations does not apply to pressure control problems which occur during tripping operations. As a result, when pressure control problems occur while tripping, drilling procedures are often applied, confusion reigns, and disaster results.

TRIP SHEETS AND FILLING PROCEDURES

Prior to a trip, it is assumed that the well is under control and that a trip can be made safely if the full hydrostatic is maintained. Pressure control problems which occur during a trip are generally the result of swabbing or a simple failure to keep the hole full. In either case, recognition and prevention of the problem is much easier than the cure. Accurate "trip sheets" must be kept whenever productive horizons have been penetrated or on the last trip before entering the transition zone or pay interval. The trip sheet is simply a record of the actual amount of mud used to keep the hole full while the drill string is being pulled compared to the theoretical quantity required to replace the pipe that is being removed. Properly monitoring the tripping operation and utilizing the trip sheet will forewarn the crew of potential well control problems.

In order to fill and monitor the hole properly, the drillpipe must be slugged dry with a barite pill. Difficulty in keeping the pipe dry is not an acceptable excuse for failure to fill the hole properly. If the first pill fails to dry the drillpipe, pump a second pill heavier than the first. If the pipe is dry for some time and then pulls wet, pump another pill. A common question is "How frequently should the hole be filled?" The basic factors determining frequency are regulations, critical nature of the well, and the wellbore geometry.

Often, special field rules or regulatory commissions will specify the method to be used to maintain the hydrostatic in particular fields or areas. Certainly, these rules are to be observed. It is acceptable to deviate from the established procedure with appropriate cause or when the procedure to be used is widely accepted as being more definitive than that established by regulation. Certainly, the condition of some wells is more critical than that of others. The critical nature may be related to location, depth, pressure, hydrocarbon composition or toxic nature of the formation fluids, to name a few.

To illustrate the significance of the wellbore geometry, consider Example 3.1:

Example 3.1

Given:

Well depth,	D =	10,000 feet
Mud density,	ρ =	15 ppg
Mud gradient,	ρ_m =	0.78 psi/ft
Length of stand,	L_{std} =	93 feet

Displacement of:

4 ½-inch drillpipe,	$DSP_{4\,1/2}$ =	0.525 bbl/std
2 7/8-inch drillpipe,	$DSP_{2\,7/8}$ =	0.337 bbl/std

Capacity of:

12 ¼-inch hole less pipe displacement, C_1 = 0.14012 bbl/ft

4 ¾-inch hole less pipe displacement, C_2 = 0.01829 bbl/ft

Stands of pipe to be pulled   = 10 stands

Wellbore configuration 1:

4 ½-inch drillpipe in 12 ¼-inch hole

Wellbore configuration 2:

2 7/8-inch drillpipe in 4 ¾-inch hole

Required:

Compare the loss in hydrostatic resulting from pulling 10 stands without filling from wellbore configuration 1 and wellbore configuration 2.

Solution:

Determine the displacement for 10 stands for wellbore configuration 1.

$$Displacement = (DSP_{ds})(\text{number of stands}) \qquad (3.1)$$

Where:

DSP_{ds} = Displacement of the drill string, bbls/std

$$Displacement = (0.525)(10)$$

$$Displacement = \textbf{5.25 bbls}$$

Determine hydrostatic loss for wellbore configuration 1.

$$Loss = \frac{\rho_m(Displacement)}{C_1} \qquad (3.2)$$

Where:

ρ_m = Mud gradient, psi/ft
C_1 = Hole capacity less pipe displacement, bbl/ft

$$Loss = \frac{(0.78)(5.25)}{0.14012}$$

$$Loss = \textbf{29 psi}$$

Determine the displacement for 10 stands for wellbore configuration 2.

$$Displacement = (DSP_{ds})(\text{number of stands})$$

$$Displacement = (0.337)(10)$$

$$Displacement = \textbf{3.37 bbls}$$

Determine hydrostatic loss for wellbore configuration 2.

$$Loss = \frac{\rho_m(Displacement)}{C_2}$$

$$Loss = \frac{(0.78)(3.37)}{0.01829}$$

$$Loss = \textbf{144 psi}$$

The loss in hydrostatic for the first wellbore configuration is obviously insignificant for most drilling operations. The loss of almost 150 psi in hydrostatic for the second case is much more significant. Often the trip margin, which is the difference between the mud hydrostatic and the formation pore pressure, is no more than 150 psi. Therefore, all things being equal, filling after 10 stands would be acceptable in the first instance while continuous filling would be in order in the second case.

PERIODIC FILLING PROCEDURE

Periodic filling, which is filling the hole after pulling a specified number of stands, is the minimum requirement and is usually accomplished utilizing a pump stroke counter according to a schedule. The periodic filling procedure is as follows:

Periodic Filling Procedure:

Step 1

Determine the pump capacity, C_p, bbls/stk.

Step 2

Determine the drill string displacement, DSP_{ds}, bbls/std, for each section of drill string.

Step 3

Determine the number of stands of each section of drill string to be pulled prior to filling the hole.

Step 4

Determine the theoretical number of pump strokes required to fill the hole after pulling the number of stands determined in Step 3.

Step 5

Prior to reaching the critical interval, begin maintaining a record of the number of pump strokes required to fill the hole after pulling the number of stands determined in Step 3.

Maintain the data in tabular form, comparing the number of stands pulled with the actual strokes required to fill the hole, the number of strokes theoretically required and the number of strokes required on the previous trip. This data, in tabular form, is the trip sheet. The trip sheet should be posted and maintained on the rig floor.

Step 6

Mix and pump a barite slug in order to pull the drill string dry. Wait for the hydrostatic inside the drill string to equalize.

Step 7

Pull the specified number of stands.

Step 8

Zero the stroke counter. Start the pump. Observe the return of mud to the surface and record in the appropriate column the actual number of strokes required to bring mud to the surface.

Step 9

Compare the actual number of strokes required to bring mud to the surface to the number of strokes required on the previous trip and the number of strokes theoretically determined.

The trip sheet generated by the periodic filling procedure is illustrated in Example 3.2:

Example 3.2
   Given:

Pump capacity,	C_p	=	0.1 bbl/stk
Drillpipe displacement,	DSP_{ds}	=	0.525 bbl/std
Drillpipe		=	4 ½-inch 16.60 #/ft
Drillpipe pulled between filling		=	10 stands

Actual strokes required as illustrated in Table 3.1

Strokes required on previous trip as illustrated in Table 3.1

   Required:
   Illustrate the proper trip sheet for a periodic filling procedure.

   Solution:
   Strokes per 10 stands

$$Strokes = \frac{DSP_{ds}\,(number\ of\ stands)}{C_p} \qquad (3.3)$$

   Where:
   DSP_{ds} = Displacement of the drill string, bbl/std

$$C_p \quad = \text{Pump capacity, bbl/stk}$$

$$Strokes = \frac{(0.525)(10)}{0.1}$$

Strokes = **52.5 per 10 stands**

The proper trip sheet for periodic filling is illustrated as Table 3.1:

Table 3.1
Trip Sheet
Periodic Filling Procedure

Cumulative Stands Pulled	Actual Strokes Required	Theoretical Strokes	Strokes Required on Previous Trip
10	55	52.5	56
20	58	52.5	57
30	56	52.5	56
....	....	....	....

The periodic filling procedure represents the minimum acceptable filling procedure. The "flo-sho," or drilling fluid return indicator, should not be used to indicate when circulation is established. It is preferable to zero the stroke counter, start the pump and observe the flow line returns. As a matter of practice, the displacement should be determined for each different section of the drill string. The number of stands of each section to be pulled between fillings may vary. For example, the hole should be filled after each stand of drill collars since the drill collar displacement is usually approximately five times drillpipe displacement.

CONTINUOUS FILLING PROCEDURE

In critical well situations, continuous filling is recommended using a trip tank. A trip tank is a small-volume tank (usually less than 60 barrels) which permits the discerning of fractions of a barrel. The better arrangement is with the trip tank in full view of the driller or floor crew and rigged with a small centrifugal pump for filling the tank and continuously circulating the mud inside the tank through the bell nipple or drillpipe annulus and back into the trip tank. When that mechanical arrangement is used, the hydrostatic will never drop.

The procedure would be that the hole would be filled continuously. After each 10 (or some specified number of) stands, the driller would observe, record and compare the volume pumped from the trip tank into the hole with the theoretical volume required. The procedure for continuously filling the hole using the trip tank would be as follows:

Step 1

Determine the drill string displacement, DSP_{ds}, bbls/std, for each section of drill string.

Step 2

Determine the number of stands of each section of drill string to be pulled prior to checking the trip tank.

Step 3

Determine the theoretical number of barrels required to replace the drill string pulled from the hole. Fill the hole after pulling the number of stands determined in Step 2.

Step 4

Prior to reaching the critical interval, begin maintaining a record of the number of barrels required to maintain the hydrostatic during the pulling of the number of stands determined in Step 2.

Maintain the data in tabular form, comparing the number of stands pulled with the cumulative volume required to maintain the hydrostatic, the volume required as theoretically determined and the volume required on the previous trip. This data, in tabular form, is the trip sheet and it should be posted and maintained on the rig floor.

Step 5

Mix and pump a barite slug in order to pull the drill string dry. Wait for the hydrostatic inside the drill string to equalize.

Step 6

Fill the trip tank and isolate the hole from the mud pits.

Step 7

Start the centrifugal pump and observe the return of mud to the trip tank.

Step 8

With the centrifugal pump circulating the hole, pull the number of stands specified in Step 2.

Step 9

After pulling the number of stands specified in Step 2, observe and record the number of barrels of mud transferred from the trip tank to the hole.

Step 10

Compare the number of barrels of mud transferred from the trip tank to the hole during this trip with the same volume transferred during the previous trip and the volume theoretically determined.

The trip sheet generated by the continuous filling procedure is illustrated in Example 3.3:

Example 3.3
Given:

Drillpipe		=	4 ½-inch 16.60 #/ft
Drillpipe displacement,	DSP_{ds}	=	0.525 bbl/std
Stands pulled between observations		=	10 stands

Trip tank capacity is 60 barrels in ¼-barrel increments

Actual volume required as illustrated in Table 3.2

Volume required on previous trip as illustrated in Table 3.2

Required:
Illustrate the proper trip sheet for a continuous filling procedure.

Solution:
Determine volume per 10 stands.

$$Displacement = (DSP_{ds})(\text{number of stands})$$

$$Displacement = (0.525)(10)$$

$$Displacement = \textbf{5.25 barrels per 10 stands}$$

The proper trip sheet for periodic filling is illustrated as Table 3.2:

Table 3.2
Trip Sheet
Continuous Filling Procedure

Cumulative Number of Stands	Cumulative Volume Required	Cumulative Theoretical Volume	Previous Trip
10	5.50	5.25	5.40
20	11.50	10.05	11.00
30	17.00	15.75	16.50
40	23.25	21.00	22.75
....	....	....	....

Generally, the actual volume of mud required to keep the hole full exceeds the theoretical calculations. The excess can be as much as 50%. On rare occasions, however, the actual volume requirements to keep the hole full are consistently less than those theoretically determined. Therefore, it is vitally important that the trip sheets from previous trips be kept for future reference. Whatever the fill pattern, it must be recorded faithfully for future comparison.

TRIPPING INTO THE HOLE

Not always, but in specific instances, it is prudent to monitor displacement while tripping in the hole to insure that fluid displacement is not excessive. The best means of measuring the displacement going in the hole is to displace directly into the isolated trip tank. A trip sheet exactly like Table 3.2 would be maintained. All too often crews are relaxed and not as diligent as necessary on the trip in the hole. As a result, industry history has recorded several instances where excessive displacement went unnoticed and severe pressure control problems resulted.

Calculations and experience prove that swabbing can occur while tripping out or in. Swab pressures should be calculated, as additional trip time can be more costly and hazardous than insignificant swab pressures. Swab pressures are real and should be considered. If a well is swabbed in on a trip in the hole, the influx will most probably be inside the drillpipe rather than the annulus because the frictional pressure is greater inside the drill string.

Further, the potential for problems does not disappear once the bit is on bottom. The pit level should be monitored carefully during the first circulation after reaching bottom. The evolution of the trip gas from the mud as it is circulated to the surface may reduce the total hydrostatic sufficiently to permit a kick.

Special attention is due when using inverted oil-emulsion systems. Historically, influxes into oil muds are difficult to detect. Because gas is infinitely soluble in oil, significant quantities of gas may pass undetected by the usual means until the pressure is reduced to the bubble point for that particular hydrocarbon mixture. At that time, the gas can flash out of solution, unload the annulus and result in a kick.

SHUT-IN PROCEDURE

WELL KICKS WHILE TRIPPING

When a trip sheet is maintained and the well fails to fill properly, the correct procedure is as follows:

Step 1

The hole is observed not to be filling properly. Discontinue the trip and check for flow.

Step 2

If the well is observed to be flowing, space out as may be necessary.

Step 3

Stab a full opening valve in the drillpipe and shut in the drillpipe.

Step 4

Open the choke line, close the blowout preventers and close the choke, pressure permitting.

Step 5

Observe and record the shut-in drillpipe pressure, the shut-in annulus pressure and the volume of formation fluid that has invaded the wellbore.

Step 6

Repeat Step 5 at 15-minute intervals.

Step 7

Prepare to strip or snub back to bottom.

The following is a discussion of each step:

Step 1

The hole is observed not to be filling properly. Discontinue the trip and check for flow.

If the hole is not filling properly, it should be checked for flow. The observation period is a function of experience in the area, the productivity of the productive formation, the depth of the well and the mud type. Under most conditions, 15 minutes is sufficient. In a deep well below 15,000 feet or if oil-based mud is being used, the observation period should be extended to a minimum of 30 minutes.

If the well is not observed to be flowing, the trip can be continued with the greatest caution. If a periodic filling procedure is being used, the hole should be filled after each stand and checked for flow until the operation returns to normal. If, after pulling another designated number of stands the well continues to fill improperly, the trip should be discontinued. If the well is not flowing when the trip is discontinued, the bit may be returned cautiously to bottom. In the event that the bit is returned to bottom, the displacement of each stand should be monitored closely and the well should be checked for flow after each stand.

If at any time the well is observed to be flowing, the trip should be discontinued. It is well known that in many areas of the world trips are made with the well flowing; however, these should be considered isolated instances and special cases.

Step 2

If the well is observed to be flowing, space out as may be necessary.

The drill string should be spaced out to insure that there is not a tool joint in the rams. If that is not a consideration, a tool joint is normally spotted at the connection position.

Step 3

Stab a full opening valve in the drillpipe and shut in the drillpipe.

The drillpipe should be shut in first. It is well known that the ball valves normally used to shut in the drillpipe are difficult to close under flow or pressure.

Step 4

Open the choke line, close the blowout preventers and close the choke, pressure permitting.

This step represents a soft shut-in. For a discussion of the soft shut-in as opposed to the hard shut-in, refer to the shut-in procedure for the Driller's Method in Chapter 2.

Step 5

Observe and record the shut-in drillpipe pressure, the shut-in annulus pressure and the volume of formation fluid that has invaded the wellbore.

Step 6

Repeat Step 5 at 15-minute intervals.

If the well has been swabbed in, the bubble should be below the bit, as illustrated in Figure 3.1. In that event, the shut-in drillpipe pressure will be equal to the shut-in casing pressure. The pressures and influx must be monitored at 15-minute intervals in order to insure that the well is effectively shut in, to establish the true reservoir pressure and to monitor bubble rise. Bubble rise is discussed in Chapter 4.

Step 7

Prepare to strip or snub back to bottom.

Stripping is a simple operation. However, stripping does require a means of bleeding and accurately measuring small volumes of mud as the drill string is stripped in the hole. A trip tank is adequate for measuring the mud volumes bled. In the alternative, a service company pump truck can be rigged up to the annulus and the displacement can be measured into its displacement tanks. If the gain is large and the pressures are high, stripping through the rig equipment may not be desirable. This point is more thoroughly discussed in Snubbing Operations in Chapter 6.

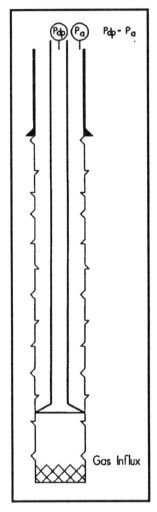

Figure 3.1 - Influx Swabbed In

Once shut in, the well is under control. Properly done, the surface pressure on the well should be less than 500 psi, which is almost insignificant. With the well shut in, there is adequate time to consider alternatives without danger of disaster. The well that has been swabbed in is a well that was under control when the bit started off bottom. Properly done, it should be a simple procedure to return the well to full control.

Once the well is shut in and under control, the options are as follows:

STRIPPING IN THE HOLE

Stripping in is not complicated if a few simple rules and concepts are observed. The procedure is as follows:

Step 1

Install a back-pressure valve on top of the safety valve in the drill string. Open the safety valve.

Step 2

Determine the displacement in barrels per stand of the drill string to be stripped into the hole. Consider that the inside of the drill string will be void.

Step 3

Determine the anticipated increase in surface pressure when the bit enters the influx according to Equation 3.4:

$$\Delta P_{incap} = \frac{DSP_{ds}}{C_{dsa}} (\rho_m - \rho_f) \tag{3.4}$$

Where:

DSP_{ds} = Displacement of drill string, bbl/std
C_{dsa} = Capacity of drill string annulus, bbl/ft
ρ_m = Mud gradient, psi/ft
ρ_f = Influx gradient, psi/ft

$$\rho_f = \frac{S_g P_b}{53.3 z_b T_b} \tag{3.5}$$

Where:

S_g = Specific gravity of gas
P_b = Bottomhole pressure, psi
T_b = Bottomhole temperature, °Rankine
z_b = Compressibility factor

Step 4

Determine the anticipated top of the influx, TOI, pursuant to Equation 3.6:

$$TOI = D - \frac{Influx\ Volume}{C_h} \tag{3.6}$$

Where:

D = Well depth, feet
C_h = Hole capacity, bbl/ft

Step 5

Prepare to lubricate the drill string with water as it passes through the surface equipment.

Step 6

Lower one stand into the hole.

Step 7

At the same time, bleed and precisely measure the displacement determined in Step 2.

Step 8

Shut in the well.

Step 9

Read and record in tabular form the shut-in casing pressure. Compare the shut-in casing pressures before and after the stand was lowered into the hole.

Note: The shut-in casing pressure should remain constant until the bit reaches the influx.

Step 10

Repeat Step 9 until the bit reaches the top of the influx as determined in Step 4.

Step 11

When the bit reaches the top of the influx as determined in Step 4, the shut-in surface pressure will increase even after the proper volume of mud is released.

Read and record the new shut-in casing pressure and compare with the original shut-in casing pressure and the anticipated increase in shut-in casing pressure as determined in Step 2.

The new shut-in casing pressure should not be greater than the original shut-in casing pressure plus the anticipated increase.

Step 12

Repeat Step 11 until the bit is on bottom.

Step 13

Once the bit is on bottom, circulate out the influx using the Driller's Method as outlined in Chapter 2.

Step 14

If necessary, circulate and raise the mud weight and trip out of the hole.

Each step will be discussed as appropriate:

Step 1

Install a back-pressure valve on top of the safety valve in the drill string. Open the safety valve.

Step 2

Determine the displacement in barrels per stand of the drill string to be stripped into the hole. Consider that the inside of the drill string will be void.

Step 3

Determine the anticipated increase in surface pressure when the bit enters the influx according to Equation 3.4:

$$\Delta P_{incap} = \frac{DSP_{ds}}{C_{dsa}}(\rho_m - \rho_f)$$

When the bit enters the influx, the influx will become longer because it will then occupy the annular area between the drill string and the hole as opposed to the open hole. Therefore, the volume of mud which is bled from the well will be replaced by the increased length of the influx. Provided that the volume of mud bled from the well is exactly as determined, the result is that the surface pressure will increase automatically by the difference between the hydrostatic of the mud and the hydrostatic of the influx, and the bottomhole pressure will remain constant. No additional influx will be permitted.

Step 4

Determine the anticipated top of the influx, TOI, pursuant to Equation 3.6:

$$TOI = D - \frac{Influx\ Volume}{C_h}$$

It is necessary to anticipate the depth at which the bit will enter the influx. As discussed, the annular pressure will suddenly increase when the bit enters the influx.

Step 5

Prepare to lubricate the drill string with water as it passes through the surface equipment.

Lubricating the string as it is lowered into the hole will reduce the weight required to cause the drill string to move. In addition, the lubricant will reduce the wear on the equipment used for stripping.

Step 6

Lower one stand into the hole.

It is important that the drill string be stripped into the hole, stand by stand.

Step 7

At the same time, bleed and precisely measure the displacement determined in Step 2.

It is vital that the volume of mud removed is exactly replaced by the drill string that is stripped into the well.

Step 8

Shut in the well.

Step 9

Read and record in tabular form the shut-in casing pressure. Compare the shut-in casing pressures before and after the stand was lowered into the hole.

Note: The shut-in casing pressure should remain constant until the bit reaches the influx.

It is important to monitor the surface pressure after each stand. Prior to the bit entering the influx, the surface pressure should remain constant. However, if the influx is gas and begins to migrate to the surface, the surface pressure will slowly begin to increase. The rate of increase in surface pressure indicates whether the increase is caused by influx migration or penetration of the influx. If the increase is due to influx penetration, the pressure will increase rapidly during the stripping of one stand. If the increase is due to influx migration, the increase is almost imperceptible for a single stand. Bubble migration and the procedure for stripping in the hole with influx migration is discussed in Chapter 4.

Step 10

Repeat Step 9 until the bit reaches the top of the influx as determined in Step 4.

Step 11

When the bit reaches the top of the influx as determined in Step 4, the shut-in surface pressure will increase even after the proper volume of mud is released.

Read and record the new shut-in casing pressure and compare with the original shut-in casing pressure and the anticipated increase in shut-in casing pressure as determined in Step 2.

The new shut-in casing pressure should not be greater than the original shut-in casing pressure plus the anticipated increase.

Step 12

Repeat Step 11 until the bit is on bottom.

Step 13

Once the bit is on bottom, circulate out the influx using the Driller's Method as outlined in Chapter 2.

Step 14

If necessary, circulate and raise the mud weight and trip out of the hole.

The procedure for stripping is illustrated in Example 3.4:

Example 3.4

Given:

Wellbore	=	Figure 3.2
Number of stands pulled	=	10 stands
Length per stand, L_{std}	=	93 ft/std
Stands stripped into the hole	=	10 stands
Drill string to be stripped	=	4 ½-inch 16.60 #/ft
Drill string displacement, DSP_{ds}	=	2 bbl/std
Mud density, ρ	=	9.6 ppg

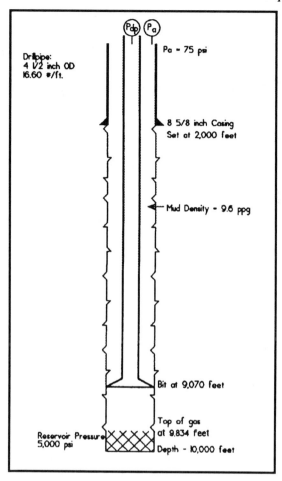

Figure 3.2 - *Wellbore Schematic*

Influx		=	10 bbls of gas
Annular capacity,	C_{dsa}	=	0.0406 bbl/ft
Depth,	D	=	10,000 feet
Hole diameter,	D_h	=	7 7/8 inches
Hole capacity,	C_h	=	0.0603 bbl/ft
Bottomhole pressure,	P_b	=	5000 psi

Bottomhole temperature, T_b =   620 °Rankine

Gas specific gravity, S_g =   0.6

Shut-in casing pressure, P_a =   75 psi

Required:
Describe the procedure for stripping the 10 stands back to bottom.

Solution:
Determine influx height, h_b :

$$h_b = \frac{Influx\ Volume}{C_h}$$ (3.7)

Where:
C_h = capacity, bbl/ft

$$h_b = \frac{10}{0.0603}$$

h_b = **166 feet**

Determine the top of the influx using Equation 3.6:

$$TOI = D - \frac{Influx\ Volume}{C_h}$$

$$TOI = 10000 - \frac{10}{0.0603}$$

TOI = **9,834 feet**

The bit will enter the influx on the **9th stand.**

Determine the depth to the bit:

$$Depth\ to\ Bit = D - (\text{number of stands})(L_{std})$$ (3.8)

Where:

D = Well depth, feet
L_{std} = Length of a stand, feet

$$Depth\ to\ Bit = 10000 - (10)(93)$$

$$Depth\ to\ Bit = \textbf{9,070 feet}$$

Determine ΔP_{incap} using Equation 3.4:

$$\rho_f = \frac{S_g(P_b)}{53.3 z_b T_b}$$

$$\rho_f = \frac{0.6(5000)}{53.3(1.1)(620)}$$

$$\rho_f = \textbf{0.0825 psi/ft}$$

$$\Delta P_{incap} = \frac{DSP_{ds}}{C_{dsa}}(\rho_m - \rho_f)$$

$$\Delta P_{incap} = \frac{2}{0.0406}(0.4992 - 0.0825)$$

$$\Delta P_{incap} = \textbf{20.53 psi/std}$$

$$\Delta P_{incap} = \textbf{0.22 psi/ft}$$

Therefore, lower 8 stands, bleeding and measuring 2 barrels with each stand. The shut-in casing pressure will remain constant at 75 psi.

Lower the 9th stand, bleeding 2 barrels. Observe that the shut-in casing pressure increases to 91 psi:

$$P_{an} = P_a + \Delta P_{incap}(feet\ of\ influx\ penetrated) \qquad (3.9)$$

Where:

P_a = Shut-in casing pressure, psi

ΔP_{incap} = Increase in pressure with bit penetration

$$P_{an} = 75 + 0.22(166 - 93)$$

$$P_{an} = \textbf{91 psi}$$

Lower the 10th stand, bleeding 2 barrels. Observe that the shut-in casing pressure increases to 112 psi.

Maintain the results as in Table 3.3:

Table 3.3 summarizes the stripping procedure. Note that if the procedure is properly done, the shut-in casing pressure remains constant until the bit penetrates the influx. This is true only if the influx is not migrating. The stripping procedure must be modified to accommodate the case of migrating influx. The proper stripping procedure, including migrating influx, is presented in Chapter 4.

Table 3.3
Stripping Procedure
Example 3.4

Stand Number	Beginning Shut-in Time	Initial Annulus Pressure	Barrels Bled	Final Shut-in Annulus Pressure
1	0800	75	2	75
2	0810	75	2	75
3	0820	75	2	75
4	0830	75	2	75
5	0840	75	2	75
6	0850	75	2	75
7	0860	75	2	75
8	0900	75	2	75
9	0910	75	2	91
10	0920	91	2	112

Example 3.4 is further summarized in Figure 3.3. Figure 3.3 illustrates the relative positions of the bit with respect to the influx during stages of the stripping operation. With 8 stands stripped into the hole, the bit is at 9,820 feet, or 14 feet above the top of the influx. Until then, the shut-in surface pressure has remained constant at 75 psi. When the 9th stand is run, the bit enters the influx and the shut-in surface pressure increases to 91 psi. On the last stand, the bit is in the influx and the shut-in surface pressure increases to 112 psi. Throughout the stripping procedure, the bottomhole pressure has remained constant at 5000 psi.

It must not be concluded that stripping is performed keeping the shut-in surface pressure constant. As illustrated in Example 3.4, had keeping the shut-in surface pressure constant been the established procedure, the well would have been underbalanced during the time that the last 2 stands were run and additional influx would have resulted. Certainly, the shut-in surface pressure must be considered. However, it is only one of the important factors.

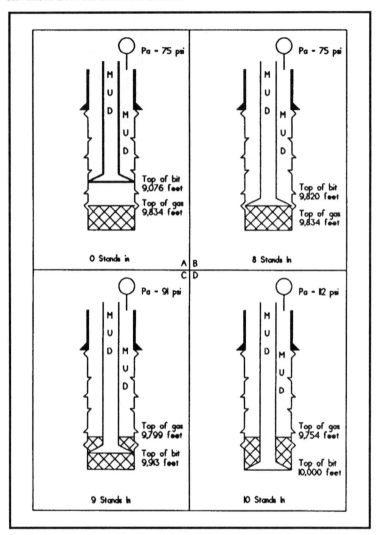

Figure 3.3

Once the bit is back on bottom, the influx can easily be circulated to the surface using the Driller's Method as outlined in Chapter 2. With the influx circulated out, the well is under control and in the same condition as before the trip began.

CHAPTER FOUR
SPECIAL CONDITIONS, PROBLEMS, and PROCEDURES IN WELL CONTROL

9 November

Took kick of approximately 100 barrels at 1530 hours. Shut in well with 2400 psi on drillpipe and casing. At 1730 pressure increased to 2700 psi. At 1930 pressure increased to 3050 psi.

10 November

0200 pressure is 2800 psi. Determined total gain was 210 barrels. At 0330 pressure is 3800 psi. Bled 10 barrels of mud and pressure dropped to 3525 psi. At 0445 pressure is 3760 psi and building 60 psi every 15 minutes. At 0545 pressure is 3980 psi and building 40 psi every 15 minutes. Pressure stabilized at 4400 psi at 1000 hours.

As illustrated in this drilling report, the circumstances surrounding a kick do not always fit the classic models. In this drilling report from southeast New Mexico, the bit was at 1,500 feet and total depth was below 14,000 feet. In addition, the surface pressures were changing rapidly. These conditions are not common to classical pressure control procedures and must be given special consideration.

This chapter is intended to discuss non-classical situations. It is important to understand classical pressure control. However, for every well control situation that fits within the classical model, there is a well control situation which bears no resemblance to the classical. According to statistics reported by the industry to the UK Health and Safety Executive, classic kicks are uncommon. For the three-year period from 1990 to 1992, of the 179 kicks reported, only 39 (22%) were classic. The student of well control must be aware of the situations in which classical procedures are appropriate and be capable of distinguishing those non-classic situations where classical procedures have no application. In addition, when the non-classic situation occurs, it is necessary to know and understand the alternatives and which has the better potential for success. In the non-classical situation, the use of classic procedures may result in the deterioration of the well condition to the point that the well is lost or the rig is burned.

SIGNIFICANCE OF SURFACE PRESSURES

In any well control situation, the pressures at the surface reflect the heart of the problem. A well out of control must obey the laws of physics and chemistry. Therefore, it is for the well control specialists to analyze and understand the problem. The well has no choice but always to accurately communicate the condition of the well. It is for us to interpret the communication properly.

A KICK IS TAKEN WHILE DRILLING

As discussed in Chapter 2 on classical pressure control, when a kick is taken while drilling and the well is shut in, the shut-in drillpipe pressure and the shut-in casing pressure are routinely recorded. The relationship between these two pressures is very important. The applicability of the classic Driller's or Wait and Weight Method must be considered in the light of the relationship between the shut-in drillpipe pressure compared with the shut-in casing pressure.

Consider the classical U-Tube Model presented as Figure 4.1. In this figure, the left side of the U-Tube represents the drillpipe while the right side represents the annulus. When the well is first shut in, the possible relationships between the shut-in drillpipe pressure, P_{dp}, and the shut-in annulus pressure, P_a, are described in Inequalities 4.1 and 4.2 and Equation 4.3 as follows:

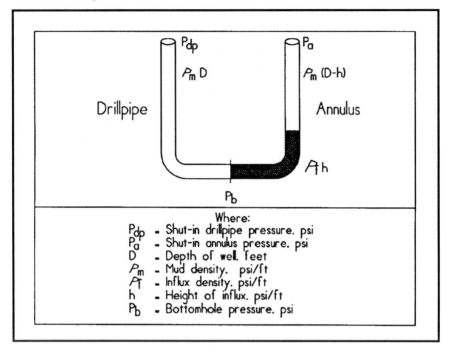

Figure 4.1 - Classic U-Tube Model

$$P_a > P_{dp} \tag{4.1}$$

$$P_a < P_{dp} \tag{4.2}$$

$$P_a \cong P_{dp} \tag{4.3}$$

With respect to the U-Tube Model in Figure 4.1, the bottomhole pressure, P_b, may be defined by the conditions on the drillpipe side and the conditions on the annulus side pursuant to Equations 2.6 and 2.7.

On the drillpipe side, P_b, is given by Equation 2.6 as follows:

$$P_b = \rho_m D + P_{dp}$$

On the annulus side, P_b, is given by Equation 2.7 as follows:

$$P_b = \rho_f h + \rho_m (D - h) + P_a$$

Equations 2.6 and 2.7 may be rearranged to give the following expressions for P_{dp} and P_a respectively:

$$P_{dp} = P_b - \rho_m D$$

and

$$P_a = P_b - \rho_m (D - h) - \rho_f h$$

To illustrate the significance of the relationship between the shut-in drillpipe pressure and the shut-in annulus pressure, it is first assumed that a kick is taken, the well is shut in and the shut-in annulus pressure is greater than the shut-in drillpipe pressure as expressed in Inequality 4.1:

$$P_a > P_{dp}$$

Substituting Equation 2.6 for the right side of the inequality and Equation 2.7 for the left side of the inequality results in the following expression:

$$P_b - \rho_m (D - h) - \rho_f h > P_b - \rho_m D$$

Expanding the terms gives

$$P_b - \rho_m D + \rho_m h - \rho_f h > P_b - \rho_m D$$

Adding $\rho_f h$ and $\rho_m D$ to both sides and subtracting P_b from both sides gives

$$P_b - P_b + \rho_m D - \rho_m D + \rho_m h + \rho_f h - \rho_f h >$$
$$P_b - P_b + \rho_m D - \rho_m D + \rho_f h$$

Simplifying the above equation results in the following:

$$\rho_m h > \rho_f h$$

Finally, dividing both sides of the inequality by h gives the consideration that

$$\rho_m > \rho_f \tag{4.4}$$

The significance of the analysis is this. When a well is on bottom drilling, a kick is taken and the well is shut in pursuant to the condition illustrated by the U-Tube Model in Figure 4.1. One of the conditions presented as expressions 4.1, 4.2 or 4.3 must describe the relationship between the shut-in drillpipe pressure and the shut-in casing pressure. For the purpose of illustration in this analysis, it was assumed that the shut-in casing pressure was greater than the shut-in drillpipe pressure as described in Inequality 4.1, which results in Inequality 4.4. Therefore, it is a certainty that, when a well kicks and is shut in, if the shut-in annulus pressure is greater than the shut-in drillpipe pressure, the density of the fluid that has entered the wellbore must be less than the density of the mud that is in the wellbore.

If the mud density is 15 ppg, the pressure on the annulus, P_a, must be greater than the pressure on the drillpipe, P_{dp}, since the heaviest naturally occurring salt water is about 9 ppg. If the mud density is 9 ppg and P_a is greater than P_{dp}, the fluid entering the wellbore is gas or some combination of gas and oil or water (determining the density of the fluid that entered the wellbore is illustrated later in this chapter).

Consider that by similar analysis it can be shown that if the Inequality 4.2 or Equation 4.3 describes the relationship between the shut-in drillpipe pressure and the shut-in annulus pressure, the following must be true:

If

$$P_a < P_{dp}$$

it must follow that

$$\rho_m < \rho_f \tag{4.5}$$

or if

$$P_a \cong P_{dp}$$

it must follow that

$$\rho_m \cong \rho_f \tag{4.6}$$

That is, if the shut-in annulus pressure is less than or equal to the shut-in drillpipe pressure, the density of the fluid which has entered the wellbore must be greater than or equal to the density of the mud in the wellbore!

Further, when the density of the drilling mud in the wellbore is greater than 10 ppg or when the influx is known to be significantly hydrocarbon, it is theoretically not possible for the shut-in casing pressure to be equal to or less than the shut-in drillpipe pressure.

Now, here is the point and it is vitally important that it be understood. If the reality is that the well is shut in, the density of the mud exceeds 10 ppg or the fluid which has entered the wellbore is known to be significantly hydrocarbon and the shut-in annulus pressure is equal to or less than the shut-in drillpipe pressure, the mathematics and reality are incompatible. When the mathematics and reality are incompatible, the mathematics have failed to describe reality. In other words, something is wrong downhole. Something is different downhole than assumed in the mathematical U-Tube Model, and that something is usually that lost circulation has occurred and the well is blowing out underground. The shut-in annulus pressure is influenced by factors other than the shut-in drillpipe pressure.

As a test, pump a small volume of mud down the drillpipe with the annulus shut in and observe the shut-in annulus pressure. No response indicates lost circulation and an underground blowout.

Whatever the cause of the incompatibility between the shut-in casing pressure and the shut-in drillpipe pressure, the significance is that under these conditions the U-Tube Model is not applicable and **CLASSIC PRESSURE CONTROL PROCEDURES ARE NOT APPLICABLE.** The Driller's Method will not control the well! The Wait and Weight Method will not control the well! "Keep the drillpipe pressure constant" has no more meaning under these conditions than any other five words in any language. "Pump standing on left foot" has as much significance and as much chance of success as "Keep the drillpipe pressure constant"!

Under these conditions, non-classical pressure control procedures must be used. There are no established procedures for non-classical pressure control operations. Each instance must be analyzed considering the unique and individual conditions, and the procedure must be detailed accordingly.

INFLUX MIGRATION

To suggest that a fluid of lesser density will migrate through a fluid of greater density should be no revelation. However, in drilling operations there are many factors that affect the rate of influx migration. In some instances, the influx has been known not to migrate.

In recent years there has been considerable research related to influx migration. In the final analysis the variables required to predict the rate of influx migration are simply not known in field operations. The old field rule of migration of approximately 1,000 feet per hour has proven to be as reliable as many much more theoretical calculations.

Some interesting and revealing observations and concepts have resulted from the research which has been conducted. Whether or not the influx will migrate depends upon the degree of mixing which occurs when the influx enters the wellbore. If the influx that enters over a relatively long period of time is significantly distributed as small bubbles in the mud and the mud is viscous, the influx may not migrate. If the influx enters in the wellbore as a continuous bubble such as is the case when the influx is swabbed into the wellbore, it will most certainly migrate. If the mud has a viscosity approaching water, the influx will most certainly migrate to the surface.

Researchers have observed many factors which will influence the rate of migration of an influx. For example, a migrating influx in a vertical annulus will travel up one side of the annulus with liquid back-flow occupying an area opposite the influx. In addition, the migrating velocity of an influx is affected by annular clearances. The smaller the annular clearances, the slower the influx will migrate. The greater the density difference between the influx and the drilling mud, the faster the influx will migrate. Therefore, the composition of the influx will affect the rate of migration as will the composition of the drilling fluid. Further, the rate of migration of an influx is reduced as the viscosity of the drilling mud is increased. Finally, an increase in the velocity of the drilling fluid will increase the migration velocity of the influx. Obviously, without specific laboratory tests on the drilling fluid, the influx fluid and the resulting mixture of the fluids in question, predictions concerning the behavior of an influx would be virtually meaningless.

As previously stated, the surface pressures are a reflection of the conditions in the wellbore. Influx migration can be observed and analyzed from the changes in the shut-in surface pressures. Basically, as the influx migrates toward the surface, the shut-in surface pressure increases provided that the geometry of the wellbore does not change. An increase in the surface pressure is the result of the reduction in the drilling mud hydrostatic above the influx as it migrates through the drilling mud toward the surface. As the influx migrates and the surface pressure increases, the pressure on the entire wellbore also increases. Thereby, the system is superpressured until the fracture gradient is exceeded or until mud is released at the surface permitting the influx to expand properly. The procedure for proper migration is discussed later in this section. At this point it is important to understand that, even under ideal conditions, the surface annular pressure will increase as the influx migrates, provided that the geometry of the wellbore does not change. If the casing is larger in the upper portion of the wellbore and the influx is permitted to expand properly, the surface pressure will decrease as the length of the influx shortens in the larger diameter casing. After decreasing as the influx enters the larger casing, the surface pressure will increase as the influx continues to migrate toward the surface.

A few field examples enlighten the points discussed. At the well in southeastern New Mexico from which the drilling report at the beginning of this chapter was excerpted, a 210-barrel influx was taken while on a trip at 14,000 feet. The top of the influx in the 6 ½-inch hole was calculated to be at 8,326 feet. The kick was taken at 1530 hours and migrated to the surface through the 11.7-ppg water-base mud at 1000 hours the following morning. The average rate of migration was 450 feet per hour.

At the Pioneer Corporation Burton in Wheeler County, Texas, all gas was routinely circulated out of the well at the end of each day. The influx would migrate through brine water from approximately 13,000 feet to the surface in eight hours. The 7-inch casing was 6 inches in internal diameter. The average rate of migration under those conditions was approximately 1,600 feet per hour.

At the E. N. Ross No. 2 near Jackson, Mississippi, a 260-barrel kick was taken inside a 7 5/8-inch liner while out of the hole on a 19,419-foot sour gas well. The top of the influx was calculated to be at 13,274 feet. A 17.4-ppg oil-base mud was being used. The initial shut-in surface pressure was 3700 psi. The pressure remained constant for the next 17 days while snubbing equipment was being rigged up, indicating that the influx did not move during that 17 days. After 17 days, the influx began to migrate into the 9 5/8-inch intermediate casing and the surface pressure declined accordingly. Six days later the influx was encountered during snubbing operations at 10,000 feet. The influx had migrated only 3,274 feet in six days. Consider Example 4.1:

Example 4.1
   Given:
   Wellbore schematic   =   Figures 4.2 and 4.3

   Top of 7 5/8-inch liner, D_l =   13,928 ft

   Well depth, D =   19,419 ft

   Influx volume   =   260 bbl

   Capacity of casing, C_{dpca} =   0.0707 bbl/ft

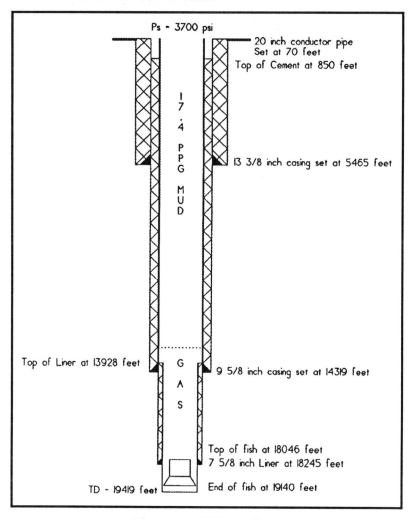

Figure 4.2 - *E.N. Ross No. 2*
Conditions after Initial Kick

Mud weight,	ρ	=	17.4 ppg OBM
Mud gradient,	ρ_m	=	0.9048 psi/ft
Bottomhole pressure,	P_b	=	16712 psi
Bottomhole temperature,	T_b	=	772 ° Rankine

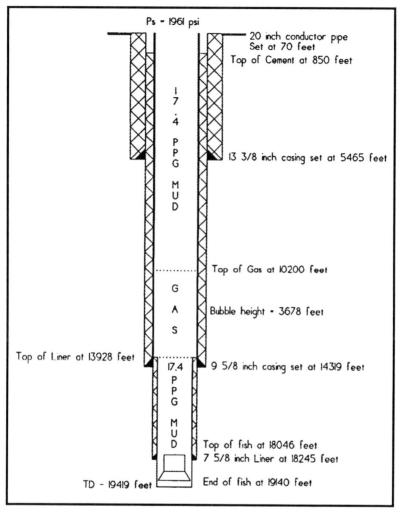

Ps - 1961 psi

20 inch conductor pipe
Set at 70 feet
Top of Cement at 850 feet

17.4 PPG MUD

13 3/8 inch casing set at 5465 feet

Top of Gas at 10200 feet

GAS

Bubble height - 3678 feet

Top of Liner at 13928 feet

17.4 PPG MUD

9 5/8 inch casing set at 14319 feet

Top of fish at 18046 feet
7 5/8 inch Liner at 18245 feet

TD - 19419 feet

End of fish at 19140 feet

Figure 4.3 *- E.N. Ross No. 2*
Bubble Migration

Temperature at 10,200 feet T_x =   650 ° Rankine

Influx gradient, ρ_f =   0.163 psi/ft on bottom

ρ_f =   0.158 psi/ft inside 9 5/8

Compressibility factor at:

10,200 feet	z_x	=	1.581
19,419 feet	z_b	=	1.988

Required:
Determine the surface pressure when the influx has migrated to 10,200 feet and is completely inside the 9 5/8-inch intermediate casing.

Solution:
From the Ideal Gas Law:

$$\frac{P_b V_b}{z_b T_b} = \frac{P_x V_x}{z_x T_x}$$

Since the influx has migrated without expansion,

$$V_b = V_x = 260 \ bbls$$

Therefore,

$$P_x = \frac{Z_x T_x P_b}{Z_b T_b}$$

or

$$P_x = \frac{16712(1.581)(650)}{(1.988)(772)}$$

$$P_x = 11190 \ psi$$

and

$$P_s = 11,190 - 0.9048(10,200)$$

$$P_s = 1961 \text{ psi}$$

Analysis of the surface pressure data was in good agreement with the actual condition encountered. The surface pressure at the time that the influx was encountered at 10,200 feet was 2000 psi. The calculated surface pressure under the given conditions was 1961 psi. It is important to note that this influx was not expanded as it migrated and the surface pressure decreased significantly. Instinctively it is anticipated that the surface pressure will increase significantly as an unexpanded influx migrates. However, under these unusual conditions, the opposite was true.

Recent research has suggested that migration analysis based upon pressure interpretations is limited due to the fact that the compressibilities of the mud, hole filter cake and formations are not routinely considered in field analyses. However, field application of the techniques described in this chapter has proven generally successful. An increase in surface pressure is the result of the reduction in drilling mud hydrostatic above the influx as the influx migrates through the drilling mud.

The concepts of influx migration and rate of migration are further illustrated in Example 4.2:

Example 4.2
   Given:

Wellbore schematic	=	Figure 4.4
Well depth, D	=	10,000 feet
Hole size, D_h	=	7 7/8 inches
Drillpipe size, D_p	=	4 ½ inches
8 5/8-inch surface casing	=	2,000 feet
Casing internal diameter, D_{ci}	=	8.017 inches

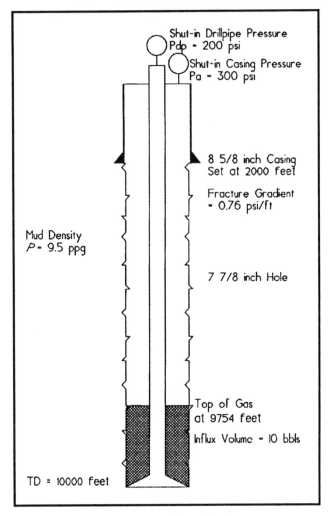

Figure 4.4 - Influx Migration

Fracture gradient,	F_g	=	0.76 psi/ft
Mud weight,	ρ	=	9.6 ppg
Mud gradient,	ρ_m	=	0.50 psi/ft

A kick is taken with the drill string on bottom and

Shut-in drillpipe pressure, P_{dp} = 200 psi

Shut-in annulus pressure, P_a = 300 psi

Pit level increase = 10 barrels

Capacity of the:

Drillpipe casing annulus, C_{dpca} = 0.0428 bbl/ft

Drillpipe hole annulus, C_{dpha} = 0.0406 bbl/ft

Depth to top of influx = 9,754 feet

Further:

The rig has lost all power and is unable to displace the influx. After one hour, the shut-in drillpipe pressure has increased to 300 psi and the shut-in annulus pressure has increased to 400 psi.

Required:
The depth to the top of the influx after one hour and the rate of influx migration

Solution:

The distance of migration, D_{mgr}, is given by Equation 4.7:

$$D_{mgr} = \frac{\Delta P_{inc}}{0.052\rho} \tag{4.7}$$

Where:

ΔP_{inc} = Pressure increase, psi

ρ = Mud weight, ppg

$$D_{mgr} = \frac{100}{(0.052)(9.6)}$$

$$D_{mgr} = \textbf{200 feet}$$

The depth to the top of the influx, D_{toi}, after one hour is

$$D_{toi} = TOI - D_{mgr} \qquad (4.8)$$

Where:

TOI = Initial top of influx, feet
D_{mgr} = Distance of migration, feet

$$D_{toi} = 9754 - 200$$

$$D_{toi} = \textbf{9,554 feet}$$

Velocity of migration, V_{mgr}, is given by

$$V_{mgr} = \frac{D_{mgr}}{Time} \qquad (4.9)$$

$$V_{mgr} = \frac{200}{1}$$

$$V_{mgr} = \textbf{200 feet per hour}$$

The condition of the well after one hour is schematically illustrated as Figure 4.5. After one hour, the shut-in surface pressures have increased by 100 psi. The shut-in drillpipe pressure has increased from 200 psi to 300 psi and the shut-in casing pressure has increased from 300 psi to 400 psi. Therefore, the drilling mud hydrostatic equivalent to

100 psi has passed from above the influx to below the influx or the influx has migrated through the equivalent of 100 psi mud hydrostatic, which is equivalent to 200 feet of mud hydrostatic. The loss in mud hydrostatic of 100 psi has been replaced by additional shut-in surface pressure of 100 psi. Therefore, the rate of migration for the first hour is 200 feet per hour.

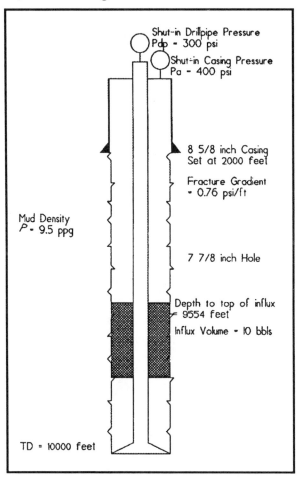

Shut-in Drillpipe Pressure
Pdp - 300 psi

Shut-in Casing Pressure
Pa - 400 psi

8 5/8 inch Casing
Set at 2000 feet

Fracture Gradient
- 0.76 psi/ft

Mud Density
P - 9.5 ppg

7 7/8 inch Hole

Depth to top of influx
- 9554 feet

Influx Volume - 10 bbls

TD = 10000 feet

Figure 4.5 - Influx Migration after One Hour

It should not be anticipated that the rate of migration will remain constant. As the influx migrates toward the surface, the velocity normally will increase. The influx will expand, the diffused bubbles will accumulate into one large bubble and the migration velocity will increase.

The influx could be permitted to migrate to the surface. The methodology would be exactly the same as the Driller's Method at 0-barrels-per-minute circulation rate. The drillpipe pressure would be kept constant by bleeding mud from the annulus. The casing pressure would have to be bled in small increments while noting the effect on the drillpipe pressure after a few seconds. For example, in this instance it would not be proper to bleed 100-psi from the annulus and wait to observe the drillpipe pressure. The proper procedure would be to permit the surface pressure to build by 100-psi increments and then to bleed the casing pressure in 25-psi increments while observing the effect on the drillpipe pressure. The exact volumes of mud bled must be measured and recorded. The drillpipe pressure must be maintained at slightly over 200 psi. In that fashion the influx could be permitted to migrate to the surface. However, once the influx reaches the surface, generally the procedure must be terminated. Bleeding influx at the surface will usually result in additional influx at the bottom of the hole.

The procedure is illustrated in Example 4.3:

Example 4.3
>   **Given:**
>   Same conditions as Example 4.2
>
>   **Required:**
>   Describe the procedure for permitting the influx to migrate to the surface.
>
>   **Solution:**
>   The effective hydrostatic of one barrel of mud, P_{hem}, in the annulus is given by Equation 4.10:

$$P_{hem} = \frac{0.052\rho}{C_{dpha}}$$

(4.10)

>   Where:
> ρ = Mud weight, ppg
> C_{dpha} = Annular capacity, bbl/ft

$$P_{hem} = \frac{0.052(9.6)}{0.0406}$$

P_{hem} = **12.3 psi/bbl**

Therefore, for each barrel of mud bled from the annulus, the minimal acceptable annulus pressure must be increased by 12.3 psi.

The following table summarizes the procedure:

Table 4.1
Procedure For Influx Migration

Time	Drillpipe Pressure	Casing Pressure	Volume Bled	Minimum Casing Pressure
0900	200	300	0.00	300
1000	300	400	0.00	300
1005	275	375	0.05	301
1010	250	350	0.10	301
1015	225	325	0.15	302
1020	200	303	0.20	303
...	...	...	...	...

As illustrated in Figure 4.6 and Table 4.1, the pressure on the surface is permitted to build to a predetermined value. This value should be calculated to consider the fracture gradient at the casing shoe in order that an underground blowout will not occur. In this first instance, the value is a 100-psi increase in surface pressure. After the surface pressure has built 100 psi to 300 psi on the drillpipe and 400 psi on the casing, the influx is expanded and the surface pressure lowered by bleeding mud from the annulus. The pressure is bled in 25-psi increments. Due to the expansion of the influx, the drillpipe pressure will return to 200 psi, but casing pressure will not return to 300 psi. Rather, the hydrostatic of the mud released from the annulus must be replaced by the equivalent pressure at the surface. In this case, 0.20 barrels were bled from the annulus and the drillpipe pressure returned to 200 psi. However, the casing pressure could not be lowered below 303 psi. The 3-psi additional casing pressure replaces the 0.20 bbl of mud hydrostatic.

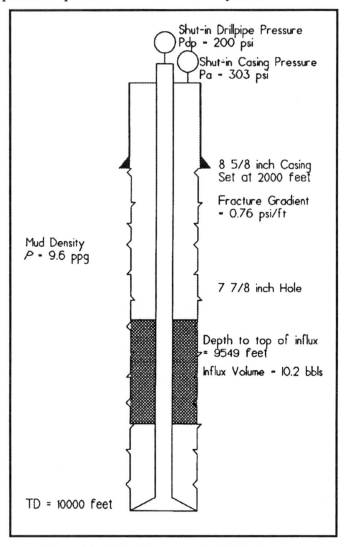

Figure 4.6 - Influx Migration after Bubble Expansion

As further illustrated in Figure 4.6, the depth to the top of the influx after the mud has been bled from the annulus is 9,549 feet. Therefore, the influx has expanded 5 feet which was the volume formerly occupied by the 0.20 barrels of mud.   Also, the influx volume has increased from 10 barrels to 10.2 barrels.

The calculations presented in this example are based on the actual theoretical calculations. In the field, the drillpipe pressure would probably be maintained at a value in excess of the original shut-in drillpipe pressure. However, the fracture gradient at the shoe must be considered in order to ensure that no underground blowout occurs.

Influx Migration — Volumetric Procedure

Influx migration without the ability to read the drillpipe pressure represents a much more difficult situation. The influx can safely be permitted to migrate to the surface if a volumetric procedure is used. Once again consider Equation 2.7:

$$P_b = \rho_f h + \rho_m (D - h) + P_a$$

Expanding Equation 2.7 gives

$$P_b = P_a + \rho_m D - \rho_m h + \rho_f h$$

The object of the procedure is to permit the influx to migrate while maintaining the bottomhole pressure constant. Therefore, the right side of Equation 2.7 must remain constant as the influx migrates. For any given conditions, $\rho_m D$, is constant.   In addition, $\rho_f h$, is constant provided the geometry of the wellbore remains constant.   To be pure theoretically, the geometry of the wellbore would have to be considered. However, to assume that the geometry is the same as on bottom is normally to err conservatively.   That is, the cross-sectional area of the annulus might increase nearer the surface, thereby reducing the hydrostatic of the influx, but it almost never decreases nearer the surface. The one obvious exception is in floating drilling operations where the influx would have to migrate through a small choke line.

Therefore, in order to permit the influx to migrate while maintaining the bottomhole pressure constant, Equation 2.7 reduces to

$$\text{Constant} = P_a + \text{Constant} - \rho_m h + \text{Constant}$$

If the bottomhole pressure is to remain constant during the influx migration, any change in the mud hydrostatic due to the expansion of the influx must be offset by a corresponding increase in the annulus pressure. In this example, if 1 barrel of mud were released from the annulus, the shut-in casing pressure could not be reduced below 312 psi. If 2 barrels of mud were released, the shut-in casing pressure could not be reduced below 324 psi.

The procedure is illustrated in Example 4.4, which is the same as Example 4.3 with the exception that the drillpipe contains a float which does not permit the shut-in drillpipe pressure to be recorded.

Example 4.4

Given:

Same conditions as Example 4.3 except the drillpipe contains a float which does not permit the shut-in drillpipe pressure to be recorded. Further, the rig has lost all power and is unable to displace the influx. After one hour, the shut-in annulus pressure has increased to 400 psi.

Required:

Describe the procedure for permitting the influx to migrate to the surface.

Solution:

The effective hydrostatic of 1 barrel of mud, P_{hem}, in the annulus is given by Equation 4.10:

$$P_{hem} = \frac{0.052\rho}{C_{dpha}}$$

$$P_{hem} = \frac{0.052(9.6)}{0.0406}$$

$$P_{hem} = 12.3 \text{ psi/bbl}$$

Therefore, for each barrel of mud bled from the annulus, the minimal acceptable annulus pressure is increased by 12.3 psi.

Maximum acceptable increase in surface pressure prior to influx expansion is equal to **100 psi.**

Bleed 1 barrel, but do not permit the surface pressure to fall below 312 psi.

After a cumulative volume of 1 barrel has been released, do not permit the surface pressure to fall below 324 psi.

After a cumulative volume of 2 barrels has been released, do not permit the surface pressure to fall below 336 psi.

After a cumulative volume of 3 barrels has been released, do not permit the surface pressure to fall below 348 psi. Continue in this manner until the influx reaches the surface.

When the influx reaches the surface, shut in the well.

The plan and instructions would be to permit the pressure to rise to a predetermined value considering the fracture gradient at the shoe. In this instance, a 100-psi increase is used. After the pressure had increased by 100 psi, mud would be released from the annulus. As much as 1 barrel would be released, provided that the shut-in casing pressure did not fall below 312 psi. Consider Table 4.2.

In this case, in the first step less than 0.20 barrels would have been bled and the casing pressure would be reduced to 312 psi. At that point, the well would be shut in and the influx permitted to migrate further up the hole.

When the shut-in surface pressure reached 412 psi, the procedure would be repeated. After a total of 1 barrel was bled from the well, the minimum casing pressure would be increased to 324 psi and the instructions would be to bleed mud to a total volume of 2 barrels, but not to permit the casing pressure to fall below 324 psi.

Table 4.2
Volumetric Procedure For Influx Migration

Time	Surface Pressure	Volume Bled	Cumulative Volume Bled	Minimum Surface Pressure
0900	300	0.00	0.00	300
1000	400	0.00	0.00	300
1005	312	0.20	0.20	312
1100	412	0.00	0.20	312
1105	312	0.20	0.40	312
1150	412	0.00	0.40	312
1155	312	0.20	0.60	312
1230	412	0.00	0.60	312
1235	312	0.25	0.85	312
1300	412	0.00	0.85	312
1305	324	0.25	1.10	324
....	....	....	....	....

When the shut-in surface pressure reached 424 psi, the procedure would be repeated. After a total of 2 barrels was bled from the well, the minimum casing pressure would be increased to 336 psi and the instructions would be to bleed mud to a total volume of 3 barrels, but not to permit the casing pressure to fall below 336 psi.

For each increment of 1 barrel of mud which was bled from the hole, the minimum shut-in surface pressure which would maintain the constant bottomhole pressure would be increased by the hydrostatic equivalent of 1 barrel of mud, which is 12.3 psi for this example. If the geometry changed as the influx migrated, the hydrostatic equivalent of 1 barrel of mud would be recalculated and the new value used.

Influx migration is a reality in well control operations and must be considered. Failure to consider the migration of the influx will usually result in unacceptable surface pressure, ruptured casing, or an underground blowout.

SAFETY FACTORS IN CLASSICAL PRESSURE CONTROL PROCEDURES

It is well established that the Driller's Method and the Wait and Weight Method are based on the classical U-Tube Model as illustrated in Figure 4.1. The displacement concept for all classical procedures regardless of the name is to determine the bottomhole pressure from the mud density and the shut-in drillpipe pressure and to keep that bottomhole pressure constant while displacing the influx. For the conditions given in Figure 4.4, the shut-in bottomhole pressure would be 5200 psi. Therefore, as illustrated in Chapter 2, the goal of the control procedure would be to circulate the influx out of the wellbore while maintaining the bottomhole pressure constant at 5200 psi.

One of the most serious and frequent well control problems encountered in the industry is the inability to bring the influx to the surface without experiencing an additional influx or causing an underground blowout. In addition, in the field difficulty is experienced starting and stopping displacement without permitting an additional influx. To address the latter problem, many have adopted "safety factor" methods.

The application of the "safety factors" arbitrarily alters the classical procedures and can result in potentially serious consequences. "Safety factors" are usually in three forms. The first is in the form of some arbitrary additional drillpipe pressure in excess of the calculated circulating pressure at the kill speed. The second is an arbitrary increase in mud density above that calculated to control the bottomhole pressure. The third is an arbitrary combination of the two. When the term "safety factor" is used, there is generally no question about the validity of the concept. Who could question "safety"? However, arbitrary "safety factors" can have serious effects on the well control procedure and can cause the very problems which they were intended to avoid! Consider Example 4.5:

Example 4.5
   Given:
      Wellbore schematic   =   Figure 4.4

      U-Tube schematic   =   Figure 4.7

Well depth, D =   10,000 feet

Hole size, D_h =   7 7/8 inches

Drillpipe size, D_p =   4 ½ inches

8 5/8-inch surface casing   =   2,000 feet

Casing internal diameter, D_{ci} =   8.017 inches

Fracture gradient, F_g =   0.76 psi/ft

Fracture pressure   =   1520 psi

Mud weight, ρ =   9.6 ppg

Mud gradient, ρ_m =   0.50 psi/ft

A kick is taken with the drill string on bottom and

Shut-in drillpipe pressure, P_{dp} =   200 psi

Shut-in annulus pressure, P_a =   300 psi

Pit level increase   =   10 barrels

Normal circulation   =   6 bpm at 60 spm

Kill rate   =   3 bpm at 30 spm

Circulating pressure at kill rate, P_{ks} =   500 psi

Pump capacity, C_p =   0.1 bbl/stk

Capacity of the:

Drillpipe casing annulus, C_{dpca} =   0.0428 bbl/ft

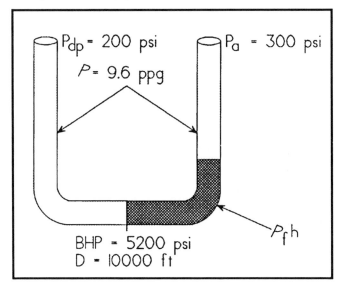

Figure 4.7 - *U-Tube Schematic*

Drillpipe hole annulus, C_{dpha} =   0.0406 bbl/ft

Initial displacement pressure, P_c =   700 psi @ 30 spm

Shut-in bottomhole pressure, P_b =   5200 psi

Maximum permissible surface pressure:

=   520 psi

Required:

1.   The consequences of adding a 200-psi "safety factor" to the initial displacement pressure in the Driller's Method.

2.   The consequences of adding a 0.5-ppg "safety factor" to the kill mud density in the Wait and Weight Method.

Solution:

1.   The consequence of adding a 200-psi "safety factor" to the initial displacement pressure is that

the pressure on the casing is increased by 200 psi to 500 psi.

The pressure at the casing shoe is increased to 1500 psi, which is perilously close to the fracture gradient. See Figure 4.8.

2.   With 10.5-ppg mud to the bit, the pressure on the left side of the U-Tube is:

$$P_{10000} = \rho_{m1}D$$

$$P_{10000} = 0.546(10000)$$

$$P_{10000} = \textbf{5460 psi}$$

Therefore, with the weighted mud at the bit, the pressure on the right side of the U-Tube is increased by **260 psi,** which would result in an underground blowout at the shoe.

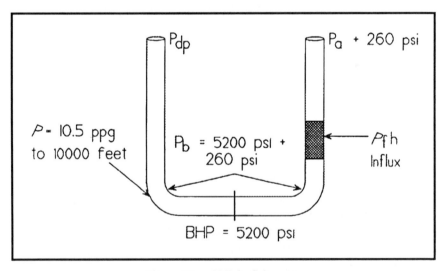

Figure 4.8 - U-Tube Schematic

In these examples, the "safety factors" were not safety factors after all. As illustrated in Figure 4.7, in the instance of a "safety factor" in the form of additional surface pressure, the additional pressure is added to the entire system. The bottomhole pressure is not being kept constant at the shut-in bottomhole pressure as intended. Rather, it is being held constant at the shut-in bottomhole pressure plus the "safety factor," which is 5400 psi in this example. To further aggravate the situation, the "safety factor" is applied to the casing shoe. Thanks to the "safety factor," the pressure at the casing shoe is increased from 1300 psi to 1500 psi, which is within 20 psi of the pressure necessary to cause an underground blowout. As the influx is circulated to the casing shoe, the pressure at the casing shoe increases. Therefore, under the conditions in Example 4.5, with the 200 psi "safety factor," an underground blowout would be inevitable!

In Figure 4.8, it is illustrated that the increase in kill-mud weight to 10.5 ppg resulted in an additional 260 psi on the entire system. The bottomhole pressure was no longer being kept constant at 5200 psi as originally conceived. It was now being kept constant at 5460 psi. This additional burden was more than the fracture gradient was capable of withstanding. By the time that the kill mud reached the bit, the annulus pressure would be well above the maximum permissible 520 psi. Therefore, under these conditions, with the additional 0.5-ppg "safety factor," an underground blowout would be inevitable.

A kill-mud density higher than calculated by classical techniques can be used, provided that Equation 2.11 is strictly adhered to:

$$P_{cn} = P_{dp} - 0.052(\rho_1 - \rho)D + \left(\frac{\rho_1}{\rho}\right)P_{ks}$$

In Equation 2.11, any additional hydrostatic pressure resulting from the increased density is subtracted from the frictional pressure. Therefore, the bottomhole pressure can be maintained constant at the calculated bottomhole pressure, which is 5200 psi in this example. Following this approach, there would be no adverse effects as a result of using the 10.5-ppg mud as opposed to the 10.0-ppg mud. Further, there would be no "safety factor" in terms of pressure at the bottom of the hole greater than the calculated shut-in bottomhole pressure. However, there would be a "safety factor" in that the pressure at the casing shoe with 10.5-ppg mud would be lower than the pressure at the casing shoe with 10.0-ppg mud. The annulus pressure profiles are further discussed in a following section. Another advantage would be that the circulating time would be less if 10.5-ppg mud were used because the trip margin would have been included in the original circulation.

A disadvantage is that the operation would fail if it became necessary to shut in the well any time after the 10.5-ppg mud reached the bit and before the influx reached the casing shoe. In that event, the pressure at the casing shoe would exceed the fracture gradient and an underground blowout would occur.

CIRCULATING A KICK OFF BOTTOM

All too often a drilling report reads like this, "Tripped out of hole. Well flowing. Tripped in with 10 stands and shut in well. Shut-in pressure 500 psi. Circulated heavy mud, keeping the drillpipe pressure constant. Shut-in pressure 5000 psi."

Attempting to circulate with the bit off bottom in a kick situation has caused as many well control situations to deteriorate as any other single operation. Simply put, **there is no classical well control procedure that applies to circulating with the bit off bottom with a formation influx in the wellbore!** The reason is that the classical U-Tube Model does not describe the wellbore condition and is not valid in this situation. If the bit is off bottom as illustrated in Figure 4.9, the U-Tube Model becomes a Y-Tube Model. The drillpipe pressure can be influenced by the operations at the choke. However, the drillpipe pressure can also be affected by the wellbore condition and activity in the bottom of the Y-Tube. It is not possible to know the relative effect of each factor.

Therefore, the concepts, technology and terminology of classical well control have no meaning or application under these circumstances. The Driller's Method is not valid. The Wait and Weight Method is not valid. Keeping the drillpipe pressure constant has no meaning. These are valid only if the U-Tube Model describes the wellbore conditions.

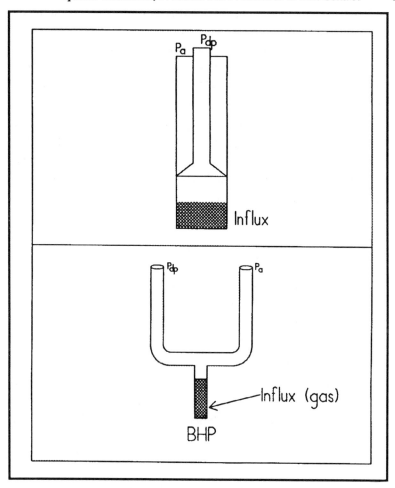

Figure 4.9 *- Circulating Off Bottom Alters U-Tube Model*

A well can be circulated safely off bottom with a kick in the hole provided that it exhibits all the characteristics of a dead well. That is, the drillpipe pressure must remain constant, the casing pressure must remain constant, the choke size must remain constant, the circulation rate must remain constant and the pit volume must remain constant. Continuing to circulate with any of these factors changing usually results in more serious well control problems.

CLASSICAL PROCEDURES — PLUGGED NOZZLE EFFECT

While a kick is being circulating out utilizing the Driller's Method or the Wait and Weight Method, it is possible for a nozzle to plug. In the event that a nozzle did plug, the choke operator would observe a sudden rise in circulating drillpipe pressure with no corresponding increase in annulus pressure. The normal reaction would be for the choke operator to open the choke in an attempt to keep the drillpipe pressure constant. Of course, when the choke is opened, the well becomes underbalanced and additional influx is permitted. Unchecked, the well will eventually unload and a blowout will follow.

A plugged nozzle does not alter the U-Tube Model. The U-Tube Model and classical pressure control procedures are still applicable. What has been altered is the frictional pressure losses in the drill string. The circulating pressure at the kill speed has been increased as a result of the plugged nozzle. The best procedure to follow when the drillpipe pressure increases suddenly is to shut in the well and restart the procedure as outlined in Chapter 2 for either the Wait and Weight Method or the Driller's Method.

CLASSICAL PROCEDURES — DRILL STRING WASHOUT EFFECT

When a washout occurs in the drill string, a loss in drillpipe pressure will be observed with no corresponding loss in annulus pressure. The only alternative is to shut in the well and analyze the problem. If the Driller's Method is being used, the analysis is simplified. As illustrated in Figure 4.10a, if the well is shut in and the influx is below the washout, the shut-in drillpipe pressure and the shut-in annulus pressure will be equal. Under these conditions, the U-Tube Model is not applicable and no classical procedure is appropriate. There are several alternatives. Probably the best general alternative is to permit the influx to migrate to the surface pursuant to the prior discussions and, once the influx has reached the surface, circulate it out. Another alternative is to locate the washout, strip out to the washout, repair the bad joint or connection, strip in the hole to bottom and resume the well control procedure.

If the Driller's Method is being utilized when the washout occurs, the well is shut in and the influx is above the washout as illustrated in Figure 4.10b, the shut-in drillpipe pressure will be less than the shut-in annulus pressure.   Under these conditions, the U-Tube Model is applicable and the influx can be circulated out by continuing the classical Driller's Method as outlined in Chapter 2. The frictional pressure losses in the drill string have been altered and the circulating pressure at the kill speed which was originally established is no longer applicable.   A new circulating pressure at the kill speed must be established as outlined in Chapter 2.   That is, hold the casing pressure constant while bringing the pump to speed.   Read the new drillpipe pressure and keep that drillpipe pressure constant while the influx is circulated to the surface.

If the Wait and Weight Method is being used, the analysis is considerably more complicated because, as illustrated in Figures 4.10c and 4.10d, the kill-weight mud has been introduced to the system. Therefore, the differences in mud hydrostatic must be included in the analysis to determine the relationship between the shut-in drillpipe pressure and the shut-in casing pressure.   Since the depth of the washout is not usually known, it may not be possible to determine a reliable relationship between the shut-in drillpipe pressure and the shut-in casing pressure.   Once the analysis is performed, the alternatives are the same as those discussed for the Driller's Method.

DETERMINATION OF SHUT-IN DRILLPIPE PRESSURES

Generally, the drillpipe pressure will stabilize within minutes after shut-in and is easily determined. In some instances, the drillpipe pressure may never build to reflect the proper bottomhole pressure, particularly in cases of long open-hole intervals at or near the fracture gradient coupled with very low productivities. When water is used as the drilling fluid, gas migration can be rapid, thereby masking the shut-in drillpipe pressure. In these instances, a good knowledge of anticipated bottomhole pressures and anticipated drillpipe pressures is beneficial in recognizing and identifying problems and providing a base for pressure control procedures.

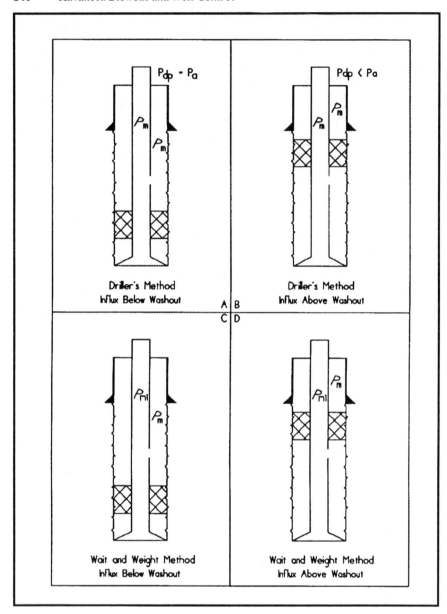

Figure 4.10

A float in the drill string complicates the determination of the drillpipe pressure; however, it can be readily determined by pumping slowly on the drillpipe and monitoring both the drillpipe and annulus pressure. When the annulus pressure first begins to increase, the drillpipe pressure at that instant is the shut-in drillpipe pressure. Another popular procedure is to pump through the float for a brief moment, holding the casing pressure constant, and then shut in with the original annulus pressure, thereby trapping the drillpipe pressure on the stand pipe gauge. Still another technique is to bring the pump to kill speed and compare the circulating pressure with the pre-recorded circulating pressure at the kill rate with the difference being the drillpipe pressure.   Still another alternative is to use a flapper-type float with a small hole drilled through the flapper which permits pressure reading but not significant flow.

DETERMINATION OF THE TYPE OF FLUID WHICH ENTERED THE WELLBORE

Of primary interest in the determination of fluid types is whether gas has entered the wellbore.   If only liquid is present, control is simplified. An accurate measurement of increase in pit level is mandatory if a reliable determination is to be made.   Example 4.6 illustrates the calculation:

Example 4.6
  Given:

Wellbore schematic		=	Figure 4.4
Well depth,	D	=	10,000 feet
Hole size,	D_h	=	7 7/8 inches
Drillpipe size,	D_p	=	4 ½ inches
8 5/8-inch surface casing		=	2,000 feet
Casing internal diameter,	D_{ci}	=	8.017 inches
Fracture gradient,	F_g	=	0.76 psi/ft
Mud weight,	ρ	=	9.6 ppg
Mud gradient,	ρ_m	=	0.50 psi/ft

A kick is taken with the drill string on bottom and

Shut-in drillpipe pressure, P_{dp} =   200 psi

Shut-in annulus pressure, P_a =   300 psi

Pit level increase   =   10 barrels

Capacity of

Drillpipe hole annulus, C_{dpha} =   0.0406 bbl/ft

Required:
Determine density of the fluid entering the wellbore.

Solution:
From Equation 2.6:

$$P_b = \rho_m D + P_{dp}$$

From Equation 2.7:

$$P_b = \rho_f h + \rho_m (D - h) + P_a$$

The height of the bubble, h, is given by Equation 3.7:

$$h = \frac{Influx\ Volume}{C_{dpha}}$$

$$h = \frac{10}{0.0406}$$

$h = 246$ feet

Solving Equations 2.6 and 2.7 simultaneously gives

$$\rho_m D + P_{dp} = \rho_f h + \rho_m (D - h) + P_a$$

The only unknown is, ρ_f; therefore, substituting and solving yields

$$\rho_f (246) = 0.5(10,000) + 200 -$$
$$0.5(10,000 - 246) - 300$$

$$\rho_f = \frac{23}{246}$$

ρ_f = **0.094 psi/ft**

Confirmation can be attained utilizing Equation 3.5:

$$\rho_f = \frac{S_g P_b}{53.3 z_b T_b}$$

$$\rho_f = \frac{0.6(5200)}{53.3(1.1)(620)}$$

ρ_f = **0.0858 psi/ft**

Therefore, since the calculated influx gradient is approximately the same as the fluid gradient determined using an assumed influx specific gravity of 0.6, the fluid in the wellbore is gas. Natural gas will usually have a fluid gradient of 0.15 psi/ft or less while brine water has a density of approximately 0.45 psi/ft and oil has a density of approximately

0.3 psi/ft. Obviously, combinations of gas, oil, and water can have a gradient anywhere between 0.1 and 0.45 psi/ft.

FRICTIONAL PRESSURE LOSSES

Frictional pressure losses inside the drill string are usually measured while frictional pressure losses in the annulus in conventional operations are usually ignored. However, in any kill operation, well-site personnel should have the means and ability to calculate the circulating pressure losses.

For laminar flow in the annulus, the frictional pressure loss is given by Equation 4.11 for the Power Law Fluid Model:

$$P_{fla} = \left[\left(\frac{2.4\bar{v}}{D_h - D_p}\right)\left(\frac{2n+1}{3n}\right)\right]^n \frac{Kl}{300(D_h - D_p)} \qquad (4.11)$$

For laminar flow inside pipe, the frictional pressure loss is given by Equation 4.12:

$$P_{fli} = \left[\left(\frac{1.6\bar{v}}{D}\right)\left(\frac{3n+1}{4n}\right)\right]^n \frac{Kl}{300} \qquad (4.12)$$

For turbulent flow in the annulus, the frictional pressure loss is given by Equation 4.13:

$$P_{fta} = \frac{7.7(10^{-5})\rho^{0.8} Q^{1.8} (PV)^{0.2} l}{(D_h - D_p)^3 (D_h + D_p)^{1.8}} \qquad (4.13)$$

For turbulent flow inside pipe, the frictional pressure loss is given by Equation 4.14:

$$P_{fti} = \frac{7.7(10^{-5})\rho^{0.8}Q^{1.8}(PV)^{0.2}l}{D_i^{4.8}}$$ (4.14)

For flow through the bit, the pressure loss is given by Equation 4.15:

$$P_{bit} = 9.14(10^{-5})\frac{\rho Q^2}{A_n^2}$$ (4.15)

Where:

P_{bit}	= Bit pressure losses, psi
P_{fla}	= Laminar annular losses, psi
P_{fta}	= Turbulent annular losses, psi
P_{fli}	= Laminar losses inside pipe, psi
P_{fti}	= Turbulent losses inside pipe, psi
$\bar{v}$	= Average velocity, fpm
D_h	= Hole diameter, inches
D_p	= Pipe outside diameter, inches
D_i	= Pipe inside diameter, inches
ρ	= Mud weight, ppg
Θ_{600}	= Viscometer reading at 600 rpm, $\dfrac{lb_f}{100\,ft^2}$
Θ_{300}	= Viscometer reading at 300 rpm, $\dfrac{lb_f}{100\,ft^2}$

$$n \quad = 3.32 \log\left(\frac{\Theta_{600}}{\Theta_{300}}\right)$$ (4.16)

$$K \quad = \frac{\Theta_{300}}{511^n}$$ (4.17)

l	= Length, feet
Q	= Volume rate of flow, gpm
PV	= Plastic viscosity, centipoise

$$= \Theta_{600} - \Theta_{300}$$ (4.18)

A_n	= Total nozzle area, inches

Flow inside the drill string is usually turbulent while flow in the annulus is normally laminar. When the flow regime is not known, make the calculations assuming both flow regimes. The calculation resulting in the greater value for frictional pressure loss is correct and defines the flow regime.

Example 4.7 illustrates a calculation:

Example 4.7
   Given:
      Wellbore schematic   =   Figure 4.4

      Well depth, D =   10,000 feet

      Hole size, D_h =   7 7/8 inches

      Drillpipe size, D_p =   4 ½ inches

      8 5/8-inch surface casing   =   2,000 feet

      Casing internal diameter, D_{ci} =   8.017 inches

      Mud weight, ρ =   9.6 ppg

      Mud gradient, ρ_m =   0.50 psi/ft

      Normal circulation rate   =   6 bpm at 60 spm

      Kill circulation rate   =   3 bpm at 30 spm

      Capacity of

      Drillpipe hole annulus, C_{dpha} =   0.0406 bbl/ft

      Θ_{600} = $25\ \dfrac{lb_f}{100\ ft^2}$

      Θ_{300} = $15\ \dfrac{lb_f}{100\ ft^2}$

Required:

The frictional pressure loss in the annulus assuming laminar flow

Solution:

$$\bar{v} = \frac{Q}{Area}$$

$$\bar{v} = \frac{126\left(\dfrac{gal}{min}\right)}{\dfrac{\pi}{4}\left(7.875^2 - 4.5^2\right)\left(in^2\right)}\left(\frac{144\dfrac{in^2}{ft^2}}{7.48\dfrac{gal}{ft^3}}\right)$$

$$\bar{v} = 74 \text{ fpm}$$

$$n = 3.32\log\left(\frac{\Theta_{600}}{\Theta_{300}}\right)$$

$$n = 3.32\log\left(\frac{25}{15}\right)$$

$$n = 0.74$$

$$K = \frac{\Theta_{300}}{511^n}$$

$$K = \frac{15}{511^{0.74}}$$

$$K = 0.15$$

$$P_{fla} = \left[\left(\frac{2.4\bar{v}}{D_h - D_p} \right) \left(\frac{2n+1}{3n} \right) \right]^n \frac{Kl}{300(D_h - D_p)}$$

$$P_{fla} = \left[\left(\frac{2.4(74)}{7.875 - 4.5} \right) \left(\frac{2(0.74)+1}{3(0.74)} \right) \right]^{0.74} \frac{(0.15)(10000)}{300(7.875 - 4.5)}$$

$$P_{fla} = 30 \text{ psi}$$

In this example, the frictional pressure loss in the annulus is only 30 psi. However, it is important to understand that the frictional pressure loss in the annulus is neglected in classical pressure control procedures. Therefore, the actual bottomhole pressure during a displacement procedure is greater than the calculated value by the value of the frictional pressure loss in the annulus. In this case, the bottomhole pressure would be held constant at 5230 psi during the Driller's Method and the Wait and Weight Method. In the final analysis, the frictional pressure loss in the annulus is a true "safety factor."

In deep wells with small annular areas, the frictional pressure loss in the annulus could be very significant and should be determined. Theoretically, if the fracture gradient at the shoe is a problem, the circulating pressure at the kill speed could be reduced by the frictional pressure loss in the annulus. For instance, using the Driller's Method in this example, the circulating pressure at the kill speed could be reduced from 700 psi at 30 spm to 670 psi at 30 spm. The bottomhole pressure would remain constant at 5200 psi, and there would be no additional influx during displacement.

ANNULUS PRESSURE PROFILES WITH CLASSICAL PROCEDURES

The annulus pressure profile as well as analysis of the pressures at the casing shoe during classic pressure control procedures provide essential insight into any well control operation.   Further, the determination of the gas volume at the surface and the time sequence for events are essential to the understanding and execution of classical pressure control procedures. Those responsible for killing the well must be informed of what is to be expected and the appropriate sequence of events. As will be illustrated in the following example, the annulus pressure will increase by almost three times during the displacement procedure and at the end dry gas will be vented for 20 minutes. Those with little experience may not expect or be mentally prepared for 20 minutes of dry gas and might be tempted to alter an otherwise sound and prudent procedure. Furthermore, confidence might be shaken by the reality of an additional 50-barrel increase in pit level. In any well control procedure, the more complete and thorough the plan, the better the chance of an expeditious and successful completion.

One of the primary problems in well control is that of circulating a kick to the surface after the well has been shut in without losing circulation and causing an underground blowout. Analysis of the annulus pressure behavior prior to initiating the displacement procedure would permit the evaluation and consideration of alternatives and probably defer a disaster.

The annulus pressure profile during classical pressure control procedures for an influx of gas can be calculated for both the Driller's Method and the Wait and Weight Method. For the Driller's Method, consider Figure 4.11. The pressure, P_x, at the top of the influx at any point in the annulus X feet from the surface is given by Equation 4.19:

$$P_x = P_a + \rho_m X \tag{4.19}$$

With the influx X feet from the surface, the surface pressure on the annulus is given by Equation 4.20:

$$P_a = P_b - \rho_m(D - h_x) - P_f \tag{4.20}$$

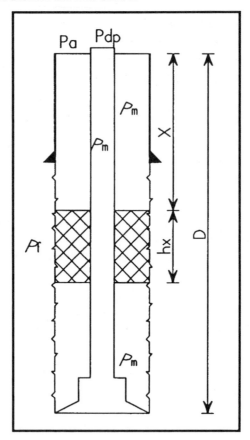

Figure 4.11 - *Wellbore Schematic for Driller's Method*

Where:

P_x	= Pressure at depth X, psi
P_a	= Annulus pressure, psi
P_b	= Bottomhole pressure, psi
ρ_m	= Mud gradient, psi/ft
D	= Well depth, feet
h_x	= Height of the influx at depth X, feet
P_f	= Pressure exerted by the influx at depth X, psi

and pursuant to the Ideal Gas Law:

$$h_x = \frac{P_b z_x T_x A_b}{P_x z_b T_b A_x} h_b$$ (4.21)

$_b$ - Denotes bottomhole conditions

$_x$ - Denotes conditions at depth X

The density of the influx is given by Equation 3.5:

$$\rho_f = \frac{S_g P_b}{53.3 z_b T_b}$$

Substituting and solving yields the Equation 4.22, which is an expression for the pressure at the top of the influx when it is any distance X from the surface when the Driller's Method is being used:

$$P_{xdm} = \frac{B}{2} + \left[\frac{B^2}{4} + \frac{P_b \rho_m z_x T_x h_b A_b}{z_b T_b A_x} \right]^{\frac{1}{2}}$$ (4.22)

$$B = P_b - \rho_m (D - X) - P_f \frac{A_b}{A_x}$$ (4.23)

Where:

$_b$ = Conditions at the bottom of the well

$_x$ = Conditions at X

X = Distance from surface to top of influx, feet

D = Depth of well, feet

P_a = Annular pressure, psi

ρ_m = Mud gradient, psi/ft

P_b = Bottomhole pressure, psi

P_f = Hydrostatic of influx, psi

z = Compressibility factor

T = Temperature, °Rankine

A = Annular area, in^2

S_g = Specific gravity of gas

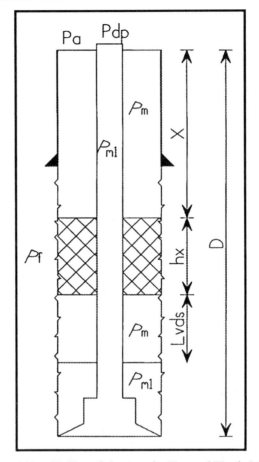

Figure 4.12 - *Wellbore Schematic for Wait and Weight Method*

Pursuant to analysis of Figure 4.12, for the Wait and Weight Method, the pressure at X is also given by Equation 4.18. However, the expression for the pressure on the annulus becomes Equation 4.24:

$$P_a = P_b - \rho_{ml}(D - X - h_x - l_{vds}) - \rho_m(l_{vds} + X) - P_f \frac{A_b}{A_x} \quad (4.24)$$

Where:

ρ_{ml} = Kill mud gradient, psi/ft
l_{vds} = Length of drill string volume in annulus, feet

Solving Equations 3.5, 4.19, 4.21, and 4.24 simultaneously results in Equation 4.25, which is an expression for the pressure at the top of the influx at any distance X from the surface when the Wait and Weight Method of displacement is being used:

$$P_{xww} = \frac{B_1}{2} + \left[\frac{B_1^2}{4} + \frac{P_b \rho_{ml} z_x T_x h_b A_b}{z_b T_b A_x} \right]^{\frac{1}{2}}$$ (4.25)

$$B_1 = P_b - \rho_{ml}(D - X) - P_f \frac{A_b}{A_x} + l_{vds}(\rho_{ml} - \rho_m)$$ (4.26)

Equation 4.22 can be used to calculate the pressure at the top of the gas bubble at any point in the annulus X distance from the surface, assuming there is no change in mud weight (Driller's Method). Similarly, Equation 4.25 can be used to calculate the pressure at the top of the gas bubble at any point in the annulus X distance from the surface, assuming that the gas bubble is displaced with weighted, ρ_{ml}, mud (Wait and Weight Method).

Depending on drill string geometry, the maximum pressure at any point in the annulus will generally occur when the bubble first reaches that point. The exception occurs when the drill collars are sufficiently larger than the drillpipe to cause a significant shortening of the influx as it passes from the drill collar annulus to the drillpipe annulus. In that instance, the pressure in the annulus will be lower than the initial shut-in annulus pressure until the influx has expanded to a length equal to its original length around the drill collars. From that point upward, the pressure in the annulus at the top of the influx will be greater than when the well was first shut in.

Example 4.8 illustrates the use of Equations 4.22 and 4.25 along with the significance and importance of the calculations:

Example 4.8
   Given:
      Wellbore schematic   =   Figure 4.4

Well depth,	D	=	10,000 feet
Hole size,	D_h	=	7 7/8 inches
Drillpipe size,	D_p	=	4 ½ inches
8 5/8-inch surface casing		=	2,000 feet
Casing internal diameter,	D_{ci}	=	8.017 inches
Fracture gradient,	F_g	=	0.76 psi/ft
Fracture pressure,		=	1520 psi
Mud weight,	ρ	=	9.6 ppg
Mud gradient,	ρ_m	=	0.50 psi/ft

A kick is taken with the drill string on bottom and

Shut-in drillpipe pressure,	P_{dp}	=	200 psi
Shut-in annulus pressure,	P_a	=	300 psi
Pit level increase		=	10 barrels
Kill-mud weight,	ρ_1	=	10 ppg
Kill-mud gradient,	ρ_{ml}	=	0.52 psi/ft
Normal circulation		=	6 bpm at 60 spm
Kill rate		=	3 bpm at 30 spm
Circulating pressure at kill rate,	P_{ks}	=	500 psi
Pump capacity,	C_p	=	0.1 bbl/stk

Capacity of the:

Drillpipe (inside),	C_{dpi} =	0.0142 bbl/ft
Drillpipe casing annulus,	C_{dpca} =	0.0428 bbl/ft
Drillpipe hole annulus,	C_{dpha} =	0.0406 bbl/ft
Initial displacement pressure,	P_c =	700 psi @ 30 spm
Shut-in bottomhole pressure,	P_b =	5200 psi

Maximum permissible surface pressure:

	=	520 psi
Ambient temperature	=	60 °F
Geothermal gradient	=	1.0 °/100 feet
$\rho_f h = P_f$	=	23 psi
h_b	=	246 feet
Annular Area,	A_b =	32.80 in^2

Required:

A.   Assuming the Driller's Method of displacement,

1.   The pressure at the casing seat when the well is first shut in.

2.   The pressure at the casing seat when the top of the gas bubble reached that point.

3.   The annulus pressure when the gas bubble first reaches the surface.

4.   The height of the gas bubble at the surface.

5.   The pressure at the casing seat when the gas bubble reaches the surface.

6.   The total pit volume increase with the influx at the surface.

7.   The surface annulus pressure profile during displacement using the Driller's Method.

B.   Using the Wait and Weight Method of displacement,

1.   The pressure at the casing seat when the well is first shut in.

2.   The pressure at the casing seat when the top of the gas bubble reached that point.

3.   The annulus pressure when the gas bubble first reaches the surface.

4.   The height of the gas bubble at the surface.

5.   The pressure at the casing seat when the gas bubble reaches the surface.

6.   The total pit volume increase with the influx at the surface.

7.   The surface annulus pressure profile during displacement using the Driller's Method.

8.   The rate at which barite must be mixed and the minimum barite required.

C.   Compare the two procedures.

D.   The significance of 600 feet of 6-inch drill collars on the annulus pressure profile.

Solution:

A.

1.   The pressure at the casing shoe at 2,000 feet when the well is first shut in is given by Equation 4.19:

$$P_x = P_a + \rho_m X$$

$$P_{2000} = 300 + 0.5(2000)$$

$$P_{2000} = \textbf{1300 psi}$$

2.   For the Driller's Method of displacement, the pressure at the casing seat at 2,000 feet when the top of the influx reaches that point is given by Equation 4.22:

$$P_{xdm} = \frac{B}{2} + \left[\frac{B^2}{4} + \frac{P_b \rho_m z_x T_x h_b A_b}{z_b T_b A_x} \right]^{\frac{1}{2}}$$

$$B = P_b - \rho_m(D - X) - P_f \frac{A_b}{A_x}$$

$$B = 5200 - 0.5(10000 - 2000) - 23 \left[\frac{32.80}{32.80} \right]$$

$$B = \textbf{1,177}$$

$$P_{2000dm} = \frac{1177}{2} +$$

$$\frac{1177^2}{4} + \frac{5200(0.5)(0.811)(540)(246)(32.80)}{(620)(1.007)(32.80)} \Bigg]^{\frac{1}{2}}$$

$$P_{2000dm} = 1480 \text{ psi}$$

3.   For the Driller's Method of displacement, the annulus pressure when the gas bubble first reaches the surface may be calculated using Equation 4.22 as follows:

$$P_{xdm} = \frac{B}{2} + \left[\frac{B^2}{4} + \frac{P_b \rho_m z_x T_x h_b A_b}{z_b T_b A_x} \right]^{\frac{1}{2}}$$

$$B = P_b - \rho_m (D - X) - P_f \frac{A_b}{A_x}$$

$$B = 5200 - 0.5(10000 - 0) - 23$$

$$B = 177$$

$$A_0 = \left(\frac{\pi}{4} \right) \left(8.017^2 - 4.5^2 \right)$$

$$A_0 = 34.58 \text{ in}^2$$

$$P_{Odm} = \frac{177}{2} +$$

$$\left[\frac{177^2}{4} + \frac{5200(0.5)(0.875)(520)(246)(32.80)}{620(1.007)(34.58)}\right]^{\frac{1}{2}}$$

$P_{Odm} = 759$ psi

4. The height of the gas bubble at the surface can be determined using Equation 4.21.

$$h_x = \frac{P_b z_x T_x A_b}{P_x z_b T_b A_x} h_b$$

$$h_0 = \frac{5200(0.875)(520)(32.80)}{759(1.007)(620)(34.58)}(246)$$

$h_0 = 1,165$ feet

5. The pressure at the casing seat when the influx reaches the surface may be calculated by adding the annulus pressure to the mud and influx hydrostatics as follows:

$$P_{2000} = P_{Odm} + P_f \frac{A_b}{A_0} + \rho_m(2000 - h_0)$$

$$P_{2000} = 759 + 23\left[\frac{32.80}{34.58}\right] + 0.50(2000 - 1165)$$

$P_{2000} = 1198$ psi

6.   The total pit volume increase with the influx at the surface. From part 4, the length of the influx when it reaches the surface is 1,165 feet.

Total pit gain = $h_0 C_{dpca}$

Total pit gain = (1165)(0.0428)

Total pit gain = **50 bbls**

7.   The surface annulus pressure profile during displacement using the Driller's Method.

Example calculation, from Part 2 with the top of the influx at 2,000 feet, the pressure at 2,000 feet is calculated to be

$P_{2000dm} = $ **1480 psi**

With the top of the influx at 2,000 feet, the surface annulus pressure is

$$P_x = P_a + \rho_m X$$

$$P_0 = 1480 - 0.50(2000)$$

$$P_0 = \textbf{480 psi}$$

Table 4.3
Surface Annulus Pressures
Driller's Method

Depth to Top of Bubble feet	Volume Pumped bbls	Annular Pressure psi
9754	0	300
9500	10	301
9000	30	303
8500	50	306
8000	69	310
7500	89	313
7000	109	318
6500	128	323
6000	148	329
5500	167	336
5000	186	346
4500	205	357
4000	224	371
3500	242	389
3000	260	412
2500	277	442
2000	294	480
1500	311	517
1000	326	579
500	340	659
0	353	759

The surface annulus pressure profile is summarized in Table 4.3 and illustrated as Figure 4.13:

B.

1.   The pressure at the casing shoe at 2,000 feet when the well is first shut in is the same for both the Driller's Method and the Wait and Weight Method:

$$P_{2000} = \textbf{1300 psi}$$

2.   For the Wait and Weight Method, the pressure at the casing seat at 2,000 feet when the top of the influx reaches that point is given by Equation 4.25:

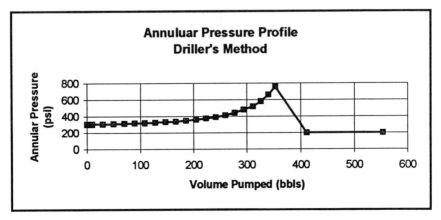

Figure 4.13

$$P_{xww} = \frac{B_1}{2} + \left[\frac{B_1^{\,2}}{4} + \frac{P_b \rho_{m1} z_x T_x h_b A_b}{z_b T_b A_x} \right]^{\frac{1}{2}}$$

$$B_1 = P_b - \rho_{m1}(D - X) -$$
$$P_f \frac{A_b}{A_x} + l_{vds}(\rho_{m1} - \rho_m)$$

$$B_1 = 5200 - 0.52(10000 - 2000) -$$
$$23\left(\frac{32.80}{32.80}\right) + \left(\frac{142}{0.0406}\right)(0.52 - 0.50)$$

$$B_1 = \textbf{1,087}$$

$$P_{2000ww} = \frac{1087}{2} +$$

$$\left[\frac{1087^2}{4} + \frac{5200(0.52)(0.816)(540)(246)(32.80)}{620(1.007)(32.80)} \right]^{\frac{1}{2}}$$

$$P_{2000ww} = 1418 \text{ psi}$$

3.   For the Wait and Weight Method of displacement, the annulus pressure when the gas bubble first reaches the surface may be calculated using Equation 4.25 as follows:

$$P_{xww} = \frac{B_1}{2} + \left[\frac{B_1^2}{4} + \frac{P_b \rho_{ml} z_x T_x h_b A_b}{z_b T_b A_x} \right]^{\frac{1}{2}}$$

$$B_1 = P_b - \rho_{ml}(D-X) - P_f \frac{A_b}{A_x} +$$

$$l_{vds}(\rho_{ml} - \rho_m)$$

$$B_1 = 5200 - 0.52(10000 - 0) - 23\left(\frac{32.80}{34.58} \right) +$$

$$\left(\frac{142}{0.0406} \right)(0.52 - 0.50)$$

$$B_1 = 48$$

$$P_{Oww} = \frac{48}{2} +$$

$$\left[\frac{48^2}{4} + \frac{5200(0.52)(0.883)(520)(246)(32.80)}{620(1.007)(34.58)} \right]^{\frac{1}{2}}$$

$P_{Oww} = $ **706 psi**

4.   The height of the gas bubble at the surface can be determined using Equation 4.21:

$$h_x = \frac{P_b z_x T_x A_b}{P_x z_b T_b A_x} h_b$$

$$h_0 = \frac{5200(0.883)(520)(32.80)}{706(1.007)(620)(34.58)}(246)$$

$h_0 = $ **1,264 feet**

5.   The pressure at the casing seat when the influx reaches the surface may be calculated by adding the annulus pressure to the mud and influx hydrostatics as follows:

$$P_{2000} = P_{Oww} + P_f \frac{A_b}{A_0} + \rho_m (2000 - h_0)$$

$$P_{2000} = 706 + 23 \left[\frac{32.80}{34.58} \right] + 0.50(2000 - 1264)$$

$P_{2000} = $ **1096 psi**

6. The total pit volume increase with the influx at the surface. From part 4, the length of the influx when it reaches the surface is 1,264 feet.

Total pit gain = $h_o C_{dpca}$

Total pit gain = (1264)(0.0428)

Total pit gain = **54 bbls**

7. The surface annulus pressure profile during displacement using the Wait and Weight Method.

Example calculation, from Part 2 with the top of the influx at 2,000 feet, the pressure at 2,000 feet is calculated to be

$P_{2000 ww} = \textbf{1418 psi}$

With the top of the influx at 2,000 feet, the surface annulus pressure is given by Equation 4.19:

$P_x = P_a + \rho_m X$

$P_0 = 1418 - 0.50(2000)$

$P_0 = \textbf{418 psi}$

The surface annulus pressure profile is summarized in Table 4.4 and illustrated as Figure 4.14:

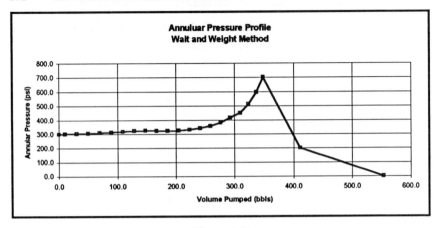

Figure 4.14

Table 4.4

Surface Annulus Pressures – Wait And Weight Method

Depth to Top of Bubble feet	Volume Pumped bbls	Annular Pressure psi
9754	0	300
9500	10	301
9000	30	303
8500	50	306
8000	69	310
7500	89	313
7000	109	318
6500	128	323
6000	148	326
5500	167	323
5000	186	323
4500	205	325
4000	223	331
3500	241	341
3000	259	358
2500	276	382
2000	293	416
1500	309	450
1000	324	513
500	337	596

8. The rate at which barite must be mixed and the minimum barite required:

$$X' = 350 \times S_m \left[\frac{W_2 - W_1}{(S_m \times 8.33) - W_2} \right] \qquad (4.27)$$

Where:

X' = Amount of barite, lbs barite/bbl mud
W_2 = Final mud weight, ppg
W_1 = Initial mud weight, ppg
S_m = Specific gravity of the weight material, water = 1

Note: S_m for barite = 4.2

$$X' = 350(4.2) \left[\frac{(10 - 9.6)}{(8.33 \times 4.2) - 10} \right]$$

X' = **23.5 lbs barite/bbl mud**

Minimum volume of mud = 548 bbls in the drillpipe and annulus:

Barite required = (23.5)(548)

Barite required = **12,878 lbs of barite**

Rate at which barite must be mixed:

Rate = (23.5)(3)

Rate = **70.5 lbs barite/minute**

C. Compare the two procedures.

Table 4.5 and Figure 4.15 compare the two methods:

Table 4.5
Comparison Of
Driller's Method
And
Wait And Weight Method

	Driller's Method	Wait and Weight Method
Pressure at casing seat when well is first shut in	1300 psi	1300 psi
Pressure at casing seat when influx reaches that point	1480 psi	1418 psi
Fracture gradient at casing seat	1520 psi	1520 psi
Annulus pressure with gas to the surface	759 psi	706 psi
Height of the gas bubble at the surface	1165 feet	1264 feet
Total pit volume increase	50 bbls	54 bbls

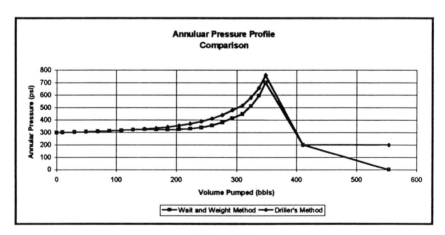

Figure 4.15

D.   The significance of 600 feet of 6-inch drill collars on the annulus pressure profile.

The significance of adding 600 feet of drill collars to each example is that the initial shut-in annulus pressure will be higher because the bubble is longer. The new annular capacity due to the additional drill collars, C_{dcha}, is 0.0253 bbls/ft. pursuant to Equation 3.7 the new influx height is:

$$h_b = \frac{Influx\ Volume}{C_{dcha}}$$

$$h_b = \frac{10}{0.0253}$$

$$h_b = \textbf{395 feet}$$

The annular pressure is given by Equation 4.20:

$$P_a = P_b - \rho_m(D - h_x) - P_f$$

$$P_a = 5200 - 0.5(10000 - 395) - 0.094(395)$$

$$P_a = \textbf{360 psi}$$

As Figures 4.13, 4.14 and 4.15 and Table 4.5 illustrate, the annulus pressure profile is exactly the same for both the Driller's Method and the Wait and Weight Method until the weighted mud reaches the bit. After the weighted mud passes the bit in the Wait and Weight Method, the annulus pressures are lower than those experienced when the Driller's Method is used. In this example, the Driller's Method may result in an underground blowout since the maximum pressure at the casing shoe, 1480 psi, is perilously close to the fracture gradient, 1520 psi. The well is more safely controlled using the Wait and Weight Method since the maximum pressure at the casing seat, 1418 psi, is almost 100 psi below the fracture gradient. Obviously, both displacement techniques would fail if a 200-psi "safety factor" was added to the circulating drillpipe pressure. This is another reason the Wait and Weight Method is preferred. In reality, the difference between the annulus pressure profiles may not be as pronounced due to influx migration during displacement. In the final analysis, the true annular pressure profile for the Wait and Weight Method is probably somewhere between the profile for the Driller's Method and the profile for the Wait and Weight Method.

In part D, with 600 feet of drill collars in the hole, the annular area is smaller. Therefore, the length of the influx will be increased from 246 feet to 395 feet. As a result, the initial shut-in annulus pressure will be 360 psi as opposed to 300 psi. Using either the Driller's Method or the Wait and Weight Method, after 5 bbls are pumped, the influx will begin to shorten as it occupies the larger volume around the drillpipe. After 15 bbls of displacement are pumped, the influx is around the drillpipe and the annulus profiles for both techniques are unaltered from that point forward (see Figure 4.16).

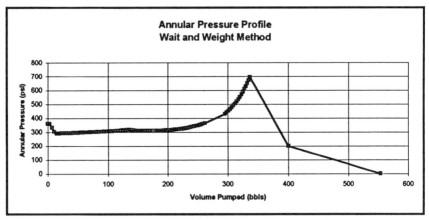

Figure 4.16

The calculations of Example 4.8 are important in order that rig personnel can be advised of the coming events. Being mentally prepared and forewarned may prevent a costly error. For example, if the Driller's

Method is used, rig personnel would expect that the maximum annulus pressure would be 759 psi, with gas to the surface; that the total pit level increase would be 50 barrels, which might cause the pits to run over; that gas would be to the surface in two hours (406 - 50)/3 or less, depending on influx migration; and that dry gas might be vented at the surface for a period of 20 minutes, at an equivalent rate of 12 million cubic feet of gas per day. Certainly, these events are enough to challenge the confidence of unsuspecting and uninitiated rig personnel and cause a major disaster. Be prepared and prepare all involved for the events to come.

CONSTANT CASING PRESSURE, CONSTANT DRILLPIPE PRESSURE MODIFICATION OF THE WAIT AND WEIGHT METHOD

After a well kick, the basic philosophy of control is the same, regardless of the procedure used. The total pressure against the kicking formation is maintained at a value sufficient to prevent further fluid entry and below the pressure that would fracture exposed formations. Simplicity is generally emphasized because decisions have to be made quickly when a well kicks, and many people are involved in the control procedure.

With the inception of current well control procedures, there was a considerable emphasis on the Driller's Method because the formation influx was immediately displaced. In recent years, it has become apparent that the Driller's Method, while simple, may result in an underground blowout that could have been prevented. Thus, many operators have adopted the Wait and Weight Method and accepted the necessity of increasing mud weight before displacement or during displacement of formation fluids. As illustrated in this chapter, the increase in mud weight requires an adjustment of drillpipe pressure during displacement of the weighted mud down the drill string. As a result, calculations are necessary to determine the pumping schedule required to maintain a constant bottomhole pressure.

To minimize the calculations, a modification of the Wait and Weight Method has become common. This procedure is known as the Constant Casing Pressure, Constant Drillpipe Pressure Wait and Weight Method. It is very simple. The only modification is that the casing pressure is kept constant at the initial shut-in casing pressure until the weighted mud reaches the bit. At that point, the drillpipe pressure is recorded and kept constant until the influx has been displaced.

The significance of this approach is illustrated by analyzing Table 4.4 and Figures 4.14 and 4.16. As illustrated, if there is no change in drill string geometry, as in the case of only one or two drill collars with heavyweight drillpipe, the casing pressure would be kept constant at 300 psi for the first 142 barrels of displacement. At that point, the casing pressure should be 352 psi. Obviously, the equivalent hydrostatic is less than the formation pressure and will continue to be less than the formation pressure throughout the displacement of the influx. Additional influx of formation fluid will be permitted and the condition of the well will deteriorate into an underground blowout.

Pursuant to Figure 4.16, if 600 feet of drill collars are present and the casing pressure is kept constant at 360 psi while the drillpipe is displaced with 142 barrels, the well will probably be safely controlled since the casing pressure at 142 barrels should be approximately 325 psi or only 35 psi less than the 360 psi being rather arbitrarily held. After the weighted mud reaches the bit, the drillpipe pressure would be held 35 psi higher than necessary to maintain the bottomhole pressure constant at 5200 psi while the influx is displaced. In this instance, that additional 35 psi would have no detrimental or harmful effects. However, each situation is unique and should be considered. For example, if larger collars are being used, the margin would be even greater.

The obvious conclusion is that the Constant Casing, Constant Drillpipe Wait and Weight Method results in arbitrary pressure profiles, which can just as easily cause deterioration of the condition of the well or loss of the well. Therefore, use of this technique is not recommended without careful consideration of the consequences which could result in the simplification being more complicated than the conventional Wait and Weight technique.

THE LOW CHOKE PRESSURE METHOD

Basically, the Low Choke Pressure Method dictates that some predetermined maximum permissible surface pressure will not be exceeded. In the event that pressure is reached, the casing pressure is maintained constant at that maximum permissible surface pressure by opening the choke, which obviously permits an additional influx of formation fluids. Once the choke size has to be reduced in order to maintain the maximum permissible annulus pressure, the drillpipe pressure is recorded and that drillpipe pressure is kept constant for the duration of the displacement procedure.

The Low Choke Pressure Method is considered by many to be a viable alternative in classic pressure control procedures when the casing pressure exceeds the maximum permissible casing pressure. However, close scrutiny dictates that this method is applicable only when the formations have low productivity.

It is understood and accepted that an additional influx will occur. However, it is assumed that the second kick will be smaller than the first. If the assumption that the second kick will be smaller than the first is correct and the second kick is in fact smaller than the first, the well might ultimately be controlled. However, if the second kick is larger than the first, the well will be lost.

In Example 4.8, the maximum permissible annulus pressure is 520 psi, which is that surface pressure which would cause fracturing at the casing seat at 2,000 feet. If the Low Choke Pressure Method was used in conjunction with the Driller's Method in Example 4.8, the condition of the well would deteriorate to an underground blowout. Consider Table 4.3 and Figure 4.13. Assuming that the Driller's Method for displacement was being used, the drillpipe pressure would be held constant at 700 psi at a pump speed of 30 strokes per minute until 311 barrels had been pumped and the casing pressure reached 520 psi. After 311 barrels of displacement, the choke would be opened in order to keep the casing pressure constant at 520 psi. As the choke was opened to maintain the casing pressure at 520 psi, the drillpipe pressure would drop below 700 psi and additional formation fluids would enter the wellbore. The well would remain underbalanced until the influx, which is approximately 406 barrels, was circulated to the surface. At a pump rate of 3 barrels per minute, the well would be underbalanced for approximately 32 minutes by as much as 186 psi!

Considering that during the original influx the underbalance was only 200 psi for much less than 32 minutes, the second influx is obviously excessive. Clearly, only a miracle will prevent an influx of formation fluids greater than the original influx of 10 barrels. That means that the well would ultimately be out of control since each successive bubble would be larger.

The Low Choke Pressure Method originated in West Texas where formations are typically high pressure and low volume. In that environment, the well is circulated on a choke until the formation depletes. In any event, the productivity is seldom sustained at more than 1 mmscfpd. Occasionally, even in that environment, the productivity is high, resulting in a serious well control problem. Therefore, in general the Low Choke Pressure Method is an acceptable procedure only in areas of known low productivity and even in that environment can result in very serious well control problems and ultimate loss of the well. When inadvertently used in areas of high productivity, disasters of major proportions can result. The Low Choke Pressure Method is not generally recommended.

REVERSE THE BUBBLE OUT THROUGH THE DRILLPIPE

It is becoming increasingly more common to reverse the influx out through the drillpipe. When the bubble is reversed out, the pressure profiles for the drillpipe and annulus are reversed, resulting in a reduction in annulus pressure when the Wait and Weight Method is used and a constant casing pressure when the Driller's Method is used. The potential hazards of bridging the annulus or plugging the bit or drillpipe are the primary objections to utilizing the reverse circulation technique; however, industry experience utilizing this technique has been successful and the industry has not experienced either drill string plugging, bit plugging or bridging in the annulus.

Planning is essential if reverse circulation is anticipated since it must be convenient to tie the drillpipe into the choke manifold system. In addition, if time permits, it is recommended that the jets be blown out of the bit. A float in the drill string is not an insurmountable obstacle in that it can be blown out of the drill string along with the jets in the bit. As an alternative, a metal bar can be pumped through the float to hold it open during the reverse circulating operation. To date, reverse circulating the influx to the surface through the drillpipe has not become a common technique; however, the limited experience of the industry to date has been good and the technical literature is promising.

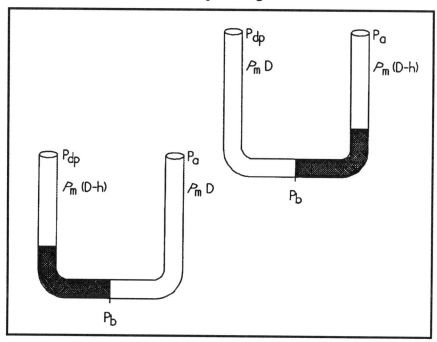

Figure 4.17 - Effect of Reverse Circulation on U-Tube Model

Operationally, reverse circulating is difficult due to the problems involved in commencing. As illustrated in Figure 4.17, the influx is in the annulus at first and within a few strokes it moves into the drillpipe. Therefore, the pressures on the drillpipe and annulus are changing rapidly during the first few strokes. The pressure on the drillpipe will increase to a value greater than the original shut-in annulus pressure while the pressure on the annulus will decline to the original shut-in drillpipe pressure. The U-Tube Model does not apply to the first few strokes; hence, there is no definitive and easy procedure for circulating the influx into the drillpipe.

Another consideration is that in classical pressure control, most of the frictional pressure losses are in the drillpipe. Therefore, the circulating pressure at the casing shoe is not affected by the frictional pressure losses. However, with reverse circulation, any frictional pressure losses in the system are applied to the casing seat, which is the weak point

in the circulating system. It is imperative that the procedure be designed
with emphasis on the fracture pressure at the casing shoe.   Consider
Example 4.9:

Example 4.9
 Given:
  Wellbore schematic   =   Figure 4.4

  Well depth, D =   10,000 feet

  Hole size, D_h =   7 7/8 inches

  Drillpipe size, D_p =   4 ½ inches

  8 5/8-inch surface casing   =   2,000 feet

  Casing internal diameter, D_{ci} =   8.017 inches

  Fracture gradient, F_g =   0.76 psi/ft

  Fracture pressure   =   1520 psi

  Mud weight, ρ =   9.6 ppg

  Mud gradient, ρ_m =   0.50 psi/ft

  A kick is taken with the drill string on bottom and

  Shut-in drillpipe pressure, P_{dp} =   200 psi

  Shut-in annulus pressure, P_a =   300 psi

  Pit level increase   =   10 barrels

  Kill mud weight, ρ_1 =   10 ppg

  Kill mud gradient, ρ_{m1} =   0.52 psi/ft

Normal circulation   =   6 bpm at 60 spm

Kill rate   =   3 bpm at 30 spm

Circulating pressure at kill rate, P_{ks} =   500 psi

Pump capacity, C_p =   0.1 bbl/stk

Capacity of the:

Drillpipe (inside), C_{dpi} =   0.0142 bbl/ft

Drillpipe casing annulus, C_{dpca} =   0.0428 bbl/ft

Drillpipe hole annulus, C_{dpha} =   0.0406 bbl/ft

Initial displacement pressure, P_c =   700 psi @ 30 spm

Shut-in bottomhole pressure, P_b =   5200 psi

Maximum permissible surface pressure:

                                               =   520 psi

Ambient temperature   =   60 °F

Geothermal gradient   =   1.0 °/100 feet

$\rho_f h = P_f$ =   23 psi

h_b =   246 feet

Required:
Prepare a schedule to reverse circulate the influx to the surface.

Solution:

1.   The first problem is to get the gain into the drillpipe.

 The maximum annulus pressure is given as = 520 psi.

 Therefore, the circulating annulus pressure must not exceed 520 psi.

2.   Determine the volume rate of flow, Q, such that the frictional pressure is less than the difference between the maximum permissible annulus pressure and the shut-in annulus pressure.

 The maximum permissible annular friction pressure is

$$520 - 300 \text{ psi} = 220 \text{ psi}$$

 From the relationship given in Equation 4.14:

$$P \propto Q^{1.8}$$

 If at Q = 30 spm and P = 500 psi, Q may be determined for P = 220 psi as follows:

$$\left(\frac{Q}{30}\right)^{1.8} = \left(\frac{220}{500}\right)$$

$$Q = 19 \text{ spm}$$

 Therefore, the bubble must be reversed at rates less than 19 spm (1.9 bpm) to prevent fracturing the shoe.

 Choose 1 barrel per minute = 10 spm.

3.   Conventionally, pump the influx up the annulus 100 strokes (an arbitrary amount) using the Driller's Method, keeping the drillpipe pressure constant.

4.   Shut in the well.

5.   Begin reverse circulation by keeping the drillpipe pressure constant at 200 psi while bringing the pump to speed approximately equal to 10 spm, with the annulus pressure not to exceed 500 psi. Total volume pumped in this step must not exceed that pumped in step 3 (100 strokes in this example).

6.   Once the rate is established, read the annulus pressure and keep that pressure constant until the influx is completely displaced.

7.   Once the influx is out, shut in and read that the drillpipe pressure equals the annulus pressure equals 200 psi.

8.   Circulating conventionally (the long way), circulate the kill-weight mud to the bit keeping the annulus pressure equal to 200 psi.

        The choke should not change during this step. The only change in pit level should be that caused by adding barite. If not, shut in.

9.   With kill mud at the bit, read the drillpipe pressure and circulate kill mud to the surface, keeping the drillpipe pressure constant.

10.   Circulate and weight up to provide desired trip margin.

Once the influx is in the drillpipe, the procedure is the same as that for the Driller's Method. Calculations such as presented in Example 4.8 can be used. With the influx inside the drillpipe, the drillpipe pressure would be 487 psi. The maximum pressure on the drillpipe would be 1308 psi and would occur when the influx reached the surface. The influx would be 2,347 feet long at the surface and the total pit gain would be 33 barrels. The time required to bring the gas to the surface would be 109 minutes and displacement of the influx would require 33 minutes. The distinct advantage is that the casing shoe would be protected from excessive pressure and an underground blowout would be avoided.

Once the influx is definitely in the drillpipe, the annulus pressure could be reduced by the difference between the original shut-in drillpipe pressure and the original shut-in annulus pressure, which is 100 psi in this example. The pressure on the annulus necessary to keep the bottomhole pressure constant at the original shut-in bottomhole pressure is the equivalent of the original shut-in drillpipe pressure plus the frictional pressure loss in the circulating system.

OVERKILL WAIT AND WEIGHT METHOD

The Overkill Wait and Weight Method is the Wait and Weight Method using a mud density greater than the calculated density for the kill-weight mud. Analysis of Example 4.8 indicates that the use of kill-weight mud to displace the gas bubble reduced the pressures in the annulus. It follows then that, if increasing the mud weight from 9.6 pounds per gallon to 10 pounds per gallon reduced the pressure at the casing seat by 62 psi as shown in Table 4.5, then additional increases in mud weight would further reduce the pressure at the casing seat. The maximum practical density would be that which would result in a vacuum on the drillpipe. The effect on the pressure at the casing seat of utilizing such a technique is calculable. Consider the utilization under the conditions described in Example 4.10:

Example 4.10
   Given:
      Example 4.6

   Required:
               1.   The effect on the pressure at the casing seat when the density of the kill-weight mud is increased to 11 ppg.

2.   The appropriate pumping schedule for displacing the influx with 11-ppg mud.

Solution:

1.   For the Wait and Weight Method, the pressure at the casing seat at 2,000 feet when the top of the influx reaches that point is given by Equation 4.25:

$$P_{xww} = \frac{B_1}{2} + \left[\frac{B_1^2}{4} + \frac{P_b P_{ml} z_x T_x h_b A_b}{z_b T_b A_x} \right]^{\frac{1}{2}}$$

$$B_1 = P_b - \rho_{ml}(D - X) - P_f \frac{A_b}{A_x}$$
$$+ l_{vds}(\rho_{ml} - \rho_m)$$

$$B_1 = 5200 - 0.572(10000 - 2000)$$
$$- 23\frac{32.80}{32.80} + \left(\frac{142}{0.0406} \right)(0.572 - 0.50)$$

$$B_1 = 852$$

$$P_{2000ww} = \frac{852}{2}$$

$$+ \left[\frac{852^2}{4} + \frac{5200(0.572)(0.828)(540)(246)(32.80)}{(620)(1.007)(32.80)} \right]^{\frac{1}{2}}$$

$$P_{2000\,ww} = 1267 \text{ psi}$$

2.   The appropriate pumping schedule may be determined using Equations 2.11 and 2.14 and is illustrated in Figure 4.18:

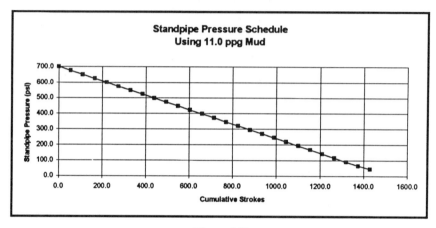

Figure 4.18

$$P_{cn} = P_{dp} - 0.052(\rho_1 - \rho)D + \left(\frac{\rho_1}{\rho}\right)P_{ks}$$

$$P_{cn} = 200 - 0.052(11.0 - 9.6)10000$$
$$+ \left(\frac{11.0}{9.6}\right)500$$

$$P_{cn} = 45 \text{ psi}$$

From Equation 2.14:

$$\frac{STKS}{25\,psi} = \frac{25(STB)}{P_c - P_{cn}}$$

$$\frac{STKS}{25} = \frac{25(1420)}{700-45}$$

$$\frac{STKS}{25} = 55$$

Table 4.6
Stand Pipe Pressure Schedule

Cumulative Strokes	Stand Pipe Pressure
0	700
55	675
110	649
165	624
220	599
275	573
330	548
385	523
440	497
495	472
550	446
605	421
660	396
715	370
770	345
825	320
880	294
935	269
990	244
1045	218
1100	193
1155	168
1210	142
1265	117
1320	91
1375	66
1430	45

Obviously, the Driller's Method and the Wait and Weight Method result in the same annulus profile while the drillpipe is being displaced since it contains unweighted mud of density ρ. In this instance, the drillpipe capacity is 142 barrels which means that both techniques have the same effect on annulus pressure for that period regardless of the density of the kill-weight mud being used in the Wait and Weight Method. As illustrated in Table 4.4, the annulus pressure at the surface after 142 barrels is approximately 325 psi. Therefore, the pressure at the shoe at 2,000 feet after 142 barrels is approximately 1325 psi. Pursuant to Example 4.10, the pressure at the casing seat when the influx reaches the casing seat is 1267 psi. Therefore, the maximum pressure at the casing seat occurred when the weighted mud reached the bit. In this example, with the 11-ppg kill-weight mud, the maximum pressure at the casing seat was almost 200 psi less than the fracture pressure of 1520 psi. The annulus pressure profile is illustrated as Figure 4.19.

In order to maintain the bottomhole pressure constant at 5200 psi, the drillpipe pressure must be reduced systematically to 45 psi and held constant throughout the remainder of the displacement procedure. Failure to consider properly reduction in drillpipe pressure resulting from the increased density is the most common cause of failure of the Overkill Wait and Weight Method. Properly utilized as illustrated in Example 4.10, the Overkill Wait and Weight Method can be a good alternative well control displacement procedure when casing shoe pressures approach fracture pressures and an underground blowout threatens.

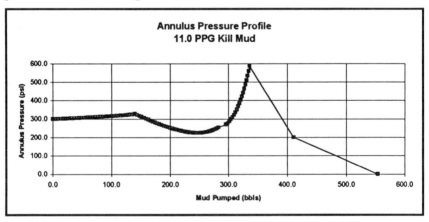

Figure 4.19

SLIM HOLE DRILLING — CONTINUOUS CORING CONSIDERATIONS

Continuous coring utilizing conventional hard-rock mining equipment to conduct slim hole drilling operations offers unique considerations in pressure control. A typical slim hole wellbore schematic is illustrated in Figure 4.20. As previously discussed, the classical pressure control displacement procedures assume that the only significant frictional pressure losses are in the drill string and that the frictional pressure losses in the annulus are negligible. As can be seen from an analysis of Figure 4.20, in slim hole drilling the conditions are reversed. That is, the frictional pressure losses in the drill string are negligible and the frictional pressure losses in the annulus are considerable. In addition, in slim hole drilling the volume of the influx is much more critical because the annulus area is small. Under normal conditions, a 1-barrel influx would not result in a significant surface annulus pressure. However, in slim hole drilling a 1-barrel influx may result in excessive annular pressures.

The best and most extensive work in this area has been done by Amoco Production Company.[1]   Sensitive flow meters have been developed which are capable of detecting extremely small influxes. Provided that the casing seats are properly selected and the influx volume is limited, classical pressure control procedures can be used. However, consideration must be given to frictional pressure losses in the annulus.

The first step is routinely to measure the surface pressure as a function of the circulation rate as discussed in Chapter 2. Modeling must then be performed to match the measured values for frictional pressure losses in order that the frictional pressure losses in the annulus may be accurately determined.   Figure 4.21 illustrates a typical pressure determination for the wellbore schematic in Figure 4.20. If, for example, the influx was to be circulated out at 40 gallons per minute, the drillpipe pressure during displacement would have to be reduced by the frictional pressure loss in the annulus which is approximately 1000 psi in Figure 4.21.   The circulating pressure at the kill speed is given by Equation 4.28:

$$P_{kssl} = P_c + P_{dp} - P_{fa}$$

$$(4.28)$$

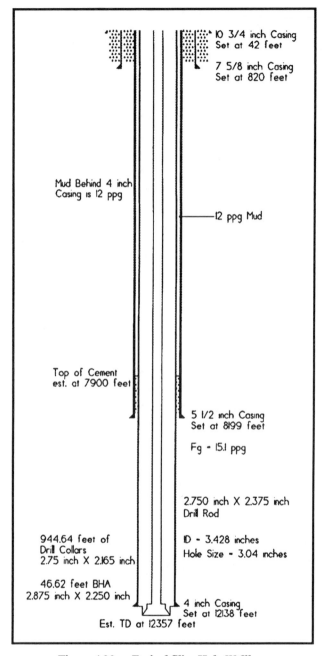

Figure 4.20 - Typical Slim Hole Wellbore

Where:

P_{kssl} = Circulating pressure at kill speed, psi

P_c = Initial circulating pressure loss, psi

P_{dp} = Shut-in drillpipe pressure, psi

P_{fa} = Frictional pressure loss in annulus, psi

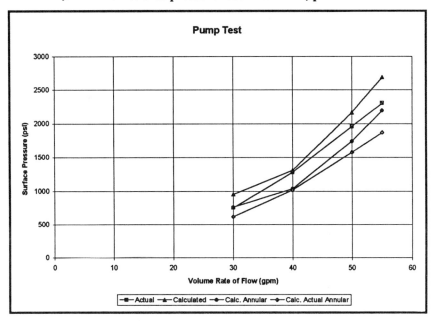

Figure 4.21

Since the influx will be displaced from the annulus before the weighted mud can reach the bit, the only classical displacement procedure applicable to most slim hole drilling operations is the Driller's Method. Therefore, pursuant to Figure 4.21 and Equation 4.28, if the shut-in drillpipe pressure was 1500 psi, the shut-in casing pressure was 1700 psi and the kill rate was 40 gpm, then the displacement pressure at the kill speed would be given by Equation 4.28 as follows:

$$P_{kssl} = P_c + P_{dp} - P_{fa}$$

$$P_{kssl} = 1300 + 1500 - 1000$$

$$P_{kssl} = \textbf{1800 psi}$$

Bringing the pump to speed for the displacement of the influx is not as straightforward as in conventional drilling operations. That is, keeping the casing pressure constant while bringing the pump to speed results in excessive pressure at the casing shoe. Therefore, the pump must be brought to speed while the casing pressure is being permitted to decline by the amount of the frictional pressure in the annulus. For example, if the shut-in casing pressure is 1700 psi, the pump would be brought to a speed of 40 gpm, permitting the casing pressure to decline by 1000 psi to 700 psi. At the same time, the drillpipe pressure would be increasing to 1800 psi. Once the appropriate drillpipe pressure was obtained, it would be held constant while the influx was circulated to the surface.

Since the annular frictional pressure losses are high and can result in an equivalent circulating density which is several pounds per gallon greater than the mud density, it is anticipated that influxes will occur when circulation is stopped for a connection. In that event, the influx can be circulated to the surface dynamically using the same circulating rate and drillpipe pressure as was used during the drilling operation.

Slim hole drilling offers the advantage of continuous coring. However, continuous coring requires the wire line retrieval of a core barrel after every 20 to 40 feet. The operation is particularly vulnerable to a kick during the retrieval of the core due to the potential swabbing action of the core barrel. The routine practice is to pump the drill rod down slowly during all core retrieval operations to insure that full hydrostatic is maintained.

The equipment required in slim hole drilling operations must be as substantial as in normal drilling operations. The choking effect of the annulus on a well flowing out of control is not as significant as might be expected. Therefore, the full compliment of well control equipment is needed. Since it is more likely that the well may flow to the surface, particular attention must be focused on the choke manifold and flare lines.

The systems outlined in Chapter 1 are applicable to slim hole drilling with two exceptions. Since a top drive is used and the drill rods have a constant outside diameter, an annular preventer is not required. In addition, since the drill rods do not have conventional upset tool joints, slip rams should be included in the blowout preventer stack to prevent the drill rods from being blown out of the hole.

STRIPPING WITH INFLUX MIGRATION

The proper procedure for stripping was discussed in Chapter 3. However, the procedure presented in Chapter 3 does not consider influx migration. To perform a stripping operation accurately with influx migration, the classical procedure in Chapter 3 must be combined with the volumetric procedures outlined in this chapter. The combined procedure is not difficult. The stripping operation is carried out normally as presented in Chapter 3. As the influx migrates, the surface pressure will slowly increase. When the surface pressure becomes unacceptable or reaches a predetermined maximum, the stripping operation is discontinued and the influx is expanded pursuant to the volumetric procedures presented in this chapter. Once the influx is expanded, the stripping operation is resumed.

Shell Oil has developed and reported a rather simple and technically correct procedure for stripping into the hole with influx migration.[2] Equipment is required which is not normally available on the rig; therefore, preplanning is a necessity. The required equipment consists of a calibrated trip tank and a calibrated stripping tank and is illustrated in Figure 4.22. Basically, the surface pressure is held constant while a stand is stripped into the hole. The mud is displaced into the calibrated trip tank. The theoretical displacement of the stand is then drained into the calibrated stripping tank. Any increase in the volume of the mud in the trip tank is recorded. The hydrostatic equivalent of the increase in volume is then added to the choke pressure. As a safety factor, the minimum annular areas are used in determining the equivalent hydrostatic of mud displaced from the hole. The procedure is as follows:

Step 1

After closing in the well, determine the influx volume and record the surface pressure.

Step 2

Determine the volume of drilling mud in the open-hole, drill collar annulus equivalent to 1 psi of mud hydrostatic.

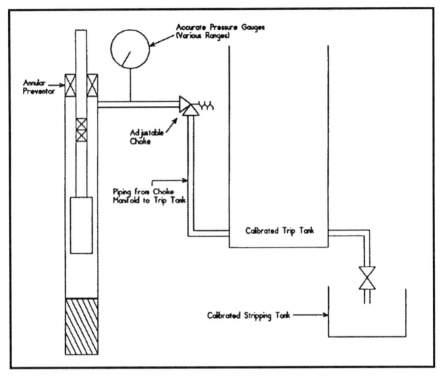

Figure 4.22 - *Stripping Equipment for Shell Method*

Step 3

Adopt a convenient working pressure increment, P_{wpi}. The working pressure increment is arbitrary, but the fracture gradient should be considered.

Step 4

Determine the volume increase in the trip tank represented by the equivalent hydrostatic of the working pressure increment.

Step 5

Determine the additional back pressure, P_{hydl}, required when the influx is in the minimum annular area, which is usually the drill collar, open-hole annulus.

Step 6

While stripping the first stand into the hole, permit the surface pressure to increase to P_{choke} where

$$P_{choke} = P_a + P_{hydl} + P_{wpi} \qquad (4.29)$$

Step 7

Strip pipe into the hole maintaining P_{choke} constant.

Step 8

After each stand is stripped into the hole, drain the theoretical displacement into the calibrated tripping tank. Record any change in the volume of mud in the trip tank.

Step 9

When the increase in volume of mud in the trip tank equals the volume determined in Step 4, increase P_{choke} by P_{wpi} and continue.

Step 10

Repeat steps 7 through 9 until the pipe is returned to bottom.

Consider Example 4.11:

Example 4.11
  Given:
    Wellbore schematic   = Figure 3.2

    Well depth, D = 10,000 feet

    Number of stands pulled   = 10 stands

    Length per stand, L_{std} = 93 ft/std

    Stands to be stripped   = 10 stands

Drillpipe to be stripped		=	4 ½ inch
Drill string displacement,	DSP_{ds}	=	2 bbl/std
Mud density,	ρ	=	9.6 ppg
Influx		=	10 bbls of gas
Capacity of			
Drillpipe annulus,	C_{dpha}	=	0.0406 bbl/ft
Hole diameter,	D_h	=	7 7/8 inches
Hole capacity,	C_h	=	0.0603 bbl/ft
Bottomhole pressure,	P_b	=	5000 psi
Bottomhole temperature,	T_b	=	620° Rankine
Gas specific gravity,	S_g	=	0.6
Shut-in casing pressure,	P_a	=	75 psi
P_{wpi}		=	50 psi
Trip tank volume		=	2 inches per barrel

Required:
Describe the procedure for stripping the 10 stands back to bottom using the Shell Method.

Solution:
Equivalent hydrostatic of 1 barrel of mud in the drillpipe annulus is given by Equation 4.10:

$$P_{hem} = \frac{0.052\rho}{C_{dpha}}$$

$$P_{hem} = \frac{0.052(9.6)}{0.0406}$$

$$P_{hem} = 12.3 \text{ psi/bbl}$$

Therefore for a 10-barrel influx:

$$P_{hydl} = 10(12.3)$$

$$P_{hydl} = 123 \text{ psi}$$

From Equation 4.29:

$$P_{choke} = 75 + 123 + 50$$

$$P_{choke} = 248 \text{ psi}$$

Therefore, strip the first stand into the hole and permit the choke pressure to increase to 250 psi. Bleed mud proportional to the amount of drillpipe stripped after reaching 250 psi.

Continue to strip pipe into the hole, keeping the surface pressure constant at 250 psi.

After each stand, bleed 2 barrels of mud into the stripping tank and record the volume in the trip tank.

When the volume in the trip tank has increased by 8 inches, increase the surface pressure to 300 psi and continue. Eight inches represents the volume of the equivalent hydrostatic of the

50 psi working pressure increment (50 psi divided by 12.3 psi/bbl multiplied by 2 inches per barrel).

OIL-BASE MUD IN PRESSURE AND WELL CONTROL OPERATIONS

The widespread use of oil-base drilling muds in deep drilling operations is relatively new. However, from the beginning, well control problems with unusual circumstances associated with oil-base muds have been observed. Typically, field personnel reports of well control problems with oil muds are as follows:

"Nothing adds up with an oil mud."

"It happened all at once! We had a 200-barrel gain in 2 minutes! There was nothing we could do! It was on us before we could do anything!"

"We saw nothing — no pit gain...nothing — until the well was flowing wildly out of control! We shut it in as fast as we could, but the pressures were too high! We lost the well!"

"We started out of the hole and it just didn't act right. It was filling OK, but it just didn't seem right. We shut it in and didn't see anything — no pressure, nothing! We still weren't satisfied; so we circulated all night. Still we saw nothing — no pit gain — no gas- cut mud — nothing. We circulated several hole volumes and just watched it. It looked OK! We pulled 10 stands, and the well began to flow. It flowed 100 barrels before we could get it shut in! I saw the pressure on the manifold go over 6000 psi before the line blew. It was all over then! We lost the rig!"

"We had just finished a trip from below 16,000 feet. Everything went great! The hole filled like it was supposed to, and the pipe went right to bottom. We had

gone back to drilling. We had already drilled our sand. We didn't drill anything new but shale and had not circulated bottoms up. Everything was going good. Then, I looked around and there was mud all over the location! I looked back toward the floor and there was mud going over the bushings! Before I could close the Hydril, the mud was going to the board! I got it shut in okay, but somehow the oil mud caught on fire in the derrick and on the rotary hose! The fire burned the rotary hose off and the well blew out up the drillpipe! It took about 30 minutes for the derrick to go! It was a terrible mess! We were paying attention! It just got us before we could do anything."

FIRE

The most obvious problem is that oil-base muds will burn. The flash point of a liquid hydrocarbon is the temperature to which it must be heated to emit sufficient flammable vapor to flash when brought into contact with a flame. The fire point of a hydrocarbon liquid is the higher temperature at which the oil vapors will continue to burn when ignited. In general, the open flash point is 50 to 70 degrees Fahrenheit less than the fire point. Most oil-base muds are made with number 2 diesel oil. The flash point for diesel is generally accepted to be about 140 degrees Fahrenheit. On that basis, the fire point would be about 200 degrees Fahrenheit. Mixing the oil-base mud with hydrocarbons from the reservoir will only increase the tendency to burn. The exposure of gas with the proper concentrations of air to any open flame or a source capable of raising the temperature of the air-gas mixture to about 1,200 degrees Fahrenheit will result in a fire.

SOLUBILITY OF NATURAL GAS IN OIL-BASE MUD

It is well known in reservoir engineering that such hydrocarbons as methane, hydrogen sulfide and carbon dioxide are extremely soluble in oil. With the popularity of oil muds used in routine drilling operations in recent years, considerable research has been performed relative to the solubility of hydrocarbons in oil muds.[3] As illustrated in Figure 4.23, the solubility of methane increases virtually linearly to approximately 6000

psi at 250 degrees Fahrenheit.   The methane solubility becomes
asymptotic at 7000 psi, which basically means that the solubility of
methane is infinite at pressures of 7000 psi or greater and temperatures of
250 degrees Fahrenheit or greater.   The pressure at which the solubility
becomes infinite is defined as the Miscibility Pressure.   Note in
Figure 4.23 that the solubility of carbon dioxide and hydrogen sulfide is
higher than that of methane.   The Miscibility Pressure of methane
decreases with temperature, as illustrated in Figure 4.24, while the
Miscibility Pressure of carbon dioxide and hydrogen sulfide increases with
temperature.   As further illustrated in Figure 4.25, the Miscibility
Pressure of methane decreases from about 8000 psi at 100 degrees to
approximately 3000 psi at 600 degrees.

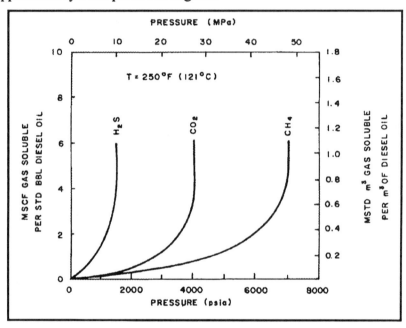

Figure 4.23 - Gas Solubility

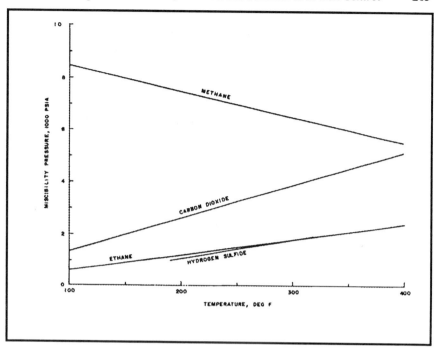

Figure 4.24

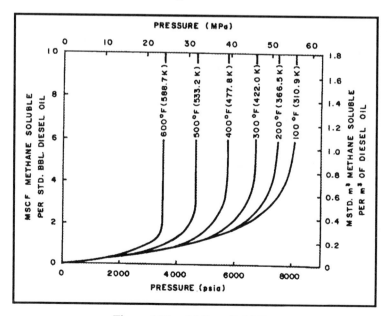

Figure 4.25 - *Methane Solubility*

The significance of this research to field operations is that in deep, high-pressure gas wells an influx of reservoir hydrocarbons can, in part, dissolve in an oil-base mud.   Another variable which will be discussed is the manner in which the influx occurs.   However, in the most simple illustration, when a kick is taken while drilling with an oil-base mud and the influx is primarily methane, the influx will dissolve into the mud system and effectively mask the presence of the influx.   That is not to say that 1 barrel of oil-base mud plus 1 barrel of reservoir hydrocarbon results in 1 barrel of a combination of the two.   However, it is certain that under the aforementioned conditions 1 barrel of oil-base mud plus 1 barrel of reservoir hydrocarbon in the gaseous phase will yield something less than 2 barrels.   Therefore, the danger signals that the man in the field normally observes are more subtle.   The rate of gain in the pit level when using an oil-mud system will be much less than the rate of gain when using a water-base system.

The exact behavior of a particular system is unpredictable.   The phase behavior of hydrocarbons is very complex and individual to the precise composition of the system.   Furthermore, the phase behavior changes as the phases change.   That is, when the gas does begin to break out of solution, the phase behavior of the remainder of the liquid phase shifts and changes.   Therefore, only generalized observations can be made.

Again, assuming the most simple example of taking a kick while drilling on bottom, the influx is partially dissolved into the oil phase of the mud system.   A typical phase diagram is illustrated in Figure 4.26.   Under such conditions, the drilling fluid is represented by point "A."   Point "A" represents a hydrocarbon system above the bubble point with all gas in solution.   As the influx is circulated up the hole, the gas will remain in solution until the bubble point is reached.   The hydrocarbon system then enters the two-phase region.   As the hydrocarbon continues up the hole, more and more gas breaks out.   As the gas breaks out, the liquid hydrostatic is replaced by the gas hydrostatic and the effective hydrostatic on bottom will decrease, permitting additional influx at an exponentially increasing rate.   This can account for the field observation of high flow rates and rapidly developing events.

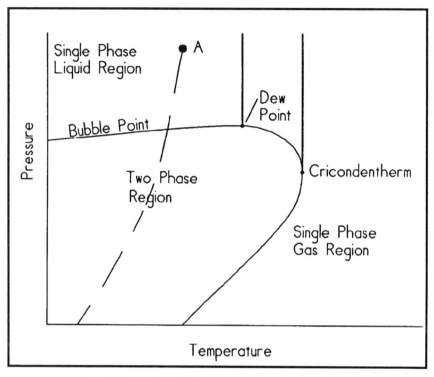

Figure 4.26 - *Typical Hydrocarbon Phase Diagram*

The pressure behavior of kicks in oil-base muds can be confusing. Using the previous illustration, if a well is shut in after observing a 10-barrel gain at the surface, it is probable that a much larger gain has been taken. However, due to the compressibility of the system, the surface pressure on the annulus may be less than that observed with a 10-barrel surface gain in a water-base mud. As the gas is circulated to the surface and begins to break out of solution, the kick will begin to behave more like a kick in a water-base mud and the annular pressure will respond accordingly. As illustrated in Figure 4.25, the solubility of methane in diesel oil is very low at low pressure and almost any reasonable temperature. Therefore, when the gas reaches the surface, the annular pressure will be much higher than expected and almost that anticipated with the same kick in a water-base mud. This scenario is schematically illustrated in Figure 4.27.

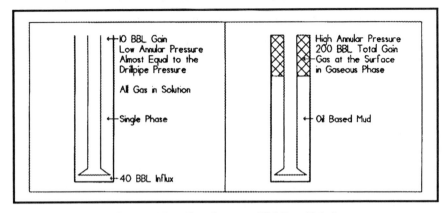

Figure 4.27 - Complexities of Oil-Based Muds

Research has shown that the solubility of the influx into the mud system is a more significant problem when the influx is widely distributed. A kick taken while on bottom drilling is an example. An influx which is in a bubble and not mixed behaves more like an influx in a water-base system. A kick which is taken while tripping or while making a connection with the pump off is an illustration.

These concepts are very difficult to visualize and verbalize. However, they are worthy of consideration. The experience of the industry along with a technical analysis dictates that much more caution must be exercised when drilling with an oil-base mud. For example, if a well is observed for flow for 15 minutes using a water base mud, it should be observed longer when an oil-base mud is being used. Unfortunately, precise field calculations cannot be made due to the complexity of the phase behavior of the hydrocarbons involved. Well control problems are significantly complicated when oil-base muds are used and the consequences are more severe. For these reasons, alternatives to oil muds should be considered in deep, high-pressure drilling.

FLOATING DRILLING AND SUBSEA OPERATION CONSIDERATIONS

SUBSEA STACK

On floating vessels the blowout preventers are located on the sea floor, necessitating redundancy on much of the equipment. Figure 4.28 is the minimum subsea stack requirement. Some operators use a double annular preventer configuration with a connector in between. The connector allows for the top package to be pulled and the top annular preventer to be repaired. The lowermost annular preventer is used as a backup when the top annular preventer fails. Shear rams, as opposed to blind rams, are a necessity in the event that conditions dictate that the drill string must be sheared and the drilling vessel moved off location. The placement of the top pipe rams above the shear rams is preferred since this arrangement allows for the hang-off string to be stripped through the pipe rams under high-pressure conditions and made up into the hang-off string. Some operators install casing rams in the upper pipe rams in a two-stack system.

The list of alternate stack arrangements is endless and it is not the intention to list the advantages and disadvantages of each. In many cases, the operator has no choice but to take the stack that comes with the rig. When a choice is available, the operator should review the potential problems in his area to determine the stack arrangement that best suits his needs, with the underlying requirement being redundancy on all critical components.

Accumulators are sometimes mounted on the subsea stack to improve response time or to activate the BOP stack acoustically in the event of an emergency when a ship is moved off location. In these instances, the precharge pressure requires adjusting to account for the hydrostatic pressure in the control lines or

$$P_{pc} = P_{inipc} + 14.7 + (l_{contline})(S_g)(0.433) \qquad (4.30)$$

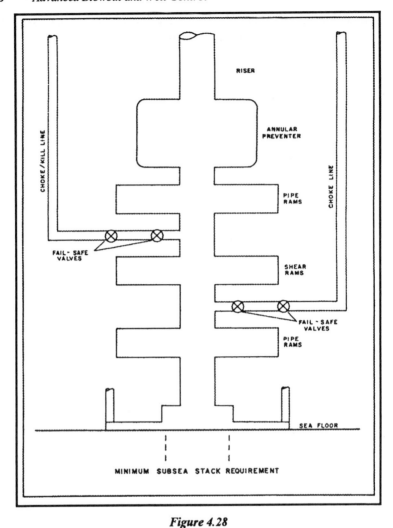

Figure 4.28

Where:

P_{pc} = Precharge pressure, psi

P_{inipc} = Initial precharge pressure, psi

$l_{contline}$ = Length of control line, feet

S_g = Specific gravity

The use of higher precharge pressure also results in a reduction in the usable volume of an accumulator bottle. It therefore follows that more bottles than normal are required. Using the Ideal Gas Law and equating volumes at precharge, minimum and maximum operating pressure:

$$\frac{P_{pc}V_{pc}}{z_{pc}T_{sc}} = \frac{P_{min}V_{min}}{z_{min}T_{ss}} = \frac{P_{max}V_{max}}{z_{min}T_{ss}}$$

Where:

ss	= Subsea
pc	= Precharge
min	= Minimum
max	= Maximum
sc	= Standard conditions
P	= Pressure, psi
V	= Volume, bbls
z	= Compressibility factor
T	= Temperature, ° Absolute

The usable volume per bottle is then $V_{min} - V_{max}$.

SPACING OUT

Due to the problems associated with pressure surge in the system and wear on the preventers associated with heave on semi-submersibles and drillships, the drill string is usually hung off on the upper pipe rams when the heave becomes excessive. The distance to the rams should be calculated after the riser has been run to ensure that the rams do not close on a tool joint. To do this, run in hole, position the tool joint 15 feet above the upper rams, close the upper rams and slowly lower the drill string until the rams take the weight of the string. Record the tide and distance to the tool joint below the rotary table for future reference.

SHUT-IN PROCEDURES

Classical shut-in procedures are modified because of the special considerations of floating drilling operations with a subsea stack. Some of the modifications are as follows:

A. While drilling

    1.   At the first indication of a kick, stop the mud pumps. Close the BOPs and, if conditions dictate, hang off as described in the previous section.

    2.   Notify the toolpusher and company representative.

    3.   Read and record the pit gain, SIDPP and SICP.

B. While tripping

    1.   At the first indication of a kick, set the slips; install and close the drillpipe safety valve.

    2.   Pick up the kelly and close the BOPs. If conditions dictate, hang off as described in the previous section.

    3.   Notify the toolpusher and company representative.

    4.   Open the safety valve; read and record the pit gain, SIDPP and SICP.

FLOATING DRILLING WELL CONTROL PROBLEMS

There are four well control problems peculiar to floating drilling operations. They are:

    1.   Fluctuations in flow rate and pit volume due to the motion of the vessel.

    2.   Friction loss in the choke line.

    3.   Reduced fracture gradient.

    4.   Gas trapped in BOP stack after circulating out a kick.

Fluctuations in Flow Rate and Pit Volume

Due to the heaving of the vessel and the related change in volume of the riser on floaters, the flow rate and pit volume fluctuate, making these primary indications of a kick difficult to interpret. Monitoring the pit volume for indications of a kick is further complicated by the pitch and roll of the vessel, as this will cause the fluid in the pits to "slosh" with the motion of the vessel — even if there is not fluid flow in or out of the pits. Many techniques have been proposed to decrease the effect of vessel movement. The pit volume totalizer, as opposed to the mechanical float, is a step in the right direction, but it requires infinite sensors to compensate totally for the entire range of vessel motion. An electronic sea-floor flow rate indicator has been devised to alleviate the problem of vessel movement. The idea is sound, but experience with this equipment is limited. For now, the industry will have to continue to monitor the surface equipment for changes in the trend. Naturally, this causes a delay in the reaction time and allows a greater influx. Knowing this, the rig personnel must be particularly alert to other kick indicators.

Frictional Loss in the Choke Line

Frictional pressure losses in the small internal-diameter (ID) choke line are negligible on land rigs but can be significant in deep-water subsea stack operations. The degree is proportional to the length and ID of the choke line. For the land rig U-Tube Dynamic Model, the bottomhole pressure, P_b, is equal to the hydrostatic of the annulus fluids, ρ_m, plus the choke back pressure, P_{ch1}:

$$P_b = \rho_m D + P_{ch1}$$

Normally, the equivalent circulating density (ECD) resulting from frictional pressure losses in the annulus is not considered since it is difficult to calculate, positive and minimal. However, in the case of a long choke line, the effects are dramatic, particularly during a start-up and shut-down operation.

With the long, small choke line the dynamic equation becomes bottomhole pressure equals the hydrostatic of the annulus fluids, ρ_m, plus the choke back pressure, P_{ch2}, plus the friction loss in the choke line, P_{fcl}:

$$P_b = \rho_m D + P_{ch2} + P_{fcl}$$

Solving the equations simultaneously results in

$$P_{ch1} = P_{ch2} + P_{fcl}$$

or

$$P_{ch2} = P_{ch1} - P_{fcl}$$

Simply said, the pressure on the choke at the surface must be reduced by the frictional pressure in the choke line. The need to understand this concept is paramount for, if the choke operator controls the back pressure to equal SICP during start up, an additional and unnecessary pressure will be imposed on the open hole formations equal to the choke line friction pressure, P_{fcl}, often with catastrophic results.

On the other hand, if the operator maintains the choke pressure constant during a shut-in operation, the choke line friction pressure is reduced to 0, reducing the bottomhole pressure by the choke line friction pressure, P_{fcl}, and allowing an additional influx.

A very simple solution and one that has gained acceptance in the industry is proposed. During any operation when the pump rate is changed, including start-up and shut-in operations, monitor the secondary choke line pressure and operate the choke to maintain this pressure constant. Since the choke line pressure loss is above the stack, then

$$P_b = \rho_m D + P_{ch\,2}$$

Therefore, monitor the secondary choke line pressure in the same manner that the primary choke back pressure (also called casing or annulus) on a land rig is monitored. The choke gauge must still be monitored for evidence of plugging.

Reduced Fracture Gradient

The classic works done in fracture gradient determination were developed for land operations and cannot be applied directly to offshore operations. The fracture gradient offshore will normally be less than an onshore gradient at equivalent depth as a result of the reduction in total overburden stress due to the air gap and sea-water gradient.

Various charts and procedures have been developed for specific areas to modify the classic methods (Eaton, Kelly and Matthews, etc.) for offshore use. The basic premise is to reduce the water depth to an equivalent section of formation by the ratio of the sea-water gradient to the overburden stress gradient at the point of interest. When using these charts, it is important to realize that some are referenced by the ratio of the subsea depth to the depth rotary table (or RKB). Consider Example 4.12:

Example 4.12
  Given:

Depth at shoe	=	2,600 feet rotary table
Water depth	=	500 feet
Air gap	=	100 feet

  Required:
  Determine the fracture gradient at the shoe.

  Solution:
  Sediment thickness = 2600 - 500 - 100 = 2,000 feet

  Fracture pressure from Figure 4.29 = 1500 psi

$$F_g = \frac{1500}{(2600)(0.052)}$$

  $F_g = 11.1 \text{ ppg}$

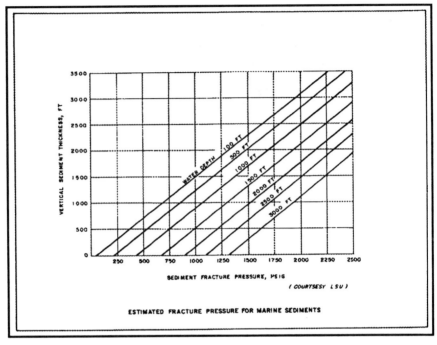

Figure 4.29

Trapped Gas after Circulating Out a Kick

Most BOP stack arrangements do not provide for the removal of the gas bubble remaining between the annular preventer and the choke line after circulating out a kick. This presents no problem in surface operations, as the pressure of this gas is minimal. However, with the use of subsea stacks, the pressure of this gas bubble is equal to the hydrostatic weight of the kill mud existing in the choke line. If improperly handled, the expansion of this bubble after opening the BOP could result in an extremely hazardous situation. Closing the lowest pipe ram to isolate the hole while displacing kill-weight mud into the riser via the kill line with the diverter closed is an acceptable method in more shallow situations. A more conservative method for deeper situations is recommended as follows:

After circulating out all kicks

1.   Close the lower pipe rams and calculate the difference in pressure between the hydrostatic head of the kill mud versus a column of water in the choke line.

2.   Displace the kill-weight mud in the BOP stack and choke line with water pumped down the kill line while holding the back pressure calculated in Step 1.

3.   Close the kill-line valve and bleed the pressure off through the gas buster.

4.   Close the diverter and open the annular preventer, allowing the remaining gas to U-tube up the choke line to the gas buster.

5.   Displace the mud in riser with kill-weight mud via kill line.

6.   Open lower pipe rams.

SHALLOW GAS KICKS

Shallow gas blowouts can be catastrophic. There is not a perfect technique for handling shallow gas blowouts. There are basically two accepted methods for handling a shallow gas kick, and the difference centers around whether the gas is diverted at the sea floor or at the surface. Operators are evenly divided between the two methods and equally dedicated to their favorite.

The disadvantage to diverting the gas at the sea floor, particularly when operating in deep water and controlling it via the choke line(s), is the additional back pressure exerted on the casing seat due to the frictional pressure drop in the choke line system.

The disadvantage to diverting the gas at the surface, particularly when operating in deep water, is the possibility of collapsing the marine riser when rapid gas expansion evacuates the riser. In addition, the gas is brought to the rig floor. Surface sands are usually unconsolidated and severe erosion is certain. Plugging and bridging are probable in some areas and can result in flow outside the casing and in cratering. Advances in diverters and diverter system design have improved surface diverting operations substantially.

References

1.   Grace, Robert D., "Further Discussion of Application of Oil Muds," SPE Drilling Engineering, September 1987, page 286.

2.   O'Bryan, P.L. and Bourgoyne, A.T., "Methods for Handling Drilled Gas in Oil Muds," SPE/IADC #16159, SPE/IADC Drilling Conference held in New Orleans, LA, 15-18 March, 1987.

3.   O'Bryan P.L. and Bourgoyne, A.T., "Swelling of Oil-based Drilling Fluids Due to Dissolved Gas," SPE 16676, presented at the 62nd Annual Technical Conference and Exhibition of the Society of Petroleum Engineers held in Dallas, TX, 27-30 September, 1987.

4..   O'Bryan, P.L. et. al., "An Experimental Study of Gas Solubility in Oil-base Drilling Fluids," SPE #15414 presented at the 61st Annual Technical Conference and Exhibition of the Society of Petroleum Engineers held in New Orleans, LA, 5-8 October, 1986.

5.   Thomas, David C., et. al., "Gas Solubility in Oil-based Drilling Fluids: Effects on Kick Detection," Journal of Petroleum Technology, June 1984, page 959.

CHAPTER FIVE
FLUID DYNAMICS IN WELL CONTROL

The use of kill fluids in well control operations is not new. However, one of the newest technical developments in well control is the engineered application of fluid dynamics. The technology of fluid dynamics is not fully utilized because most personnel involved in well control operations do not understand the engineering applications and do not have the capabilities to apply the technology at the rig in the field. The best well control procedure is the one that has predictable results from a technical as well as a mechanical perspective.

Fluid dynamics have an application in virtually every well control operation. Appropriately applied, fluids can be used cleverly to compensate for unreliable tubulars or inaccessibility. Often, when blowouts occur, the tubulars are damaged beyond expectation. For example, at a blowout in Wyoming an intermediate casing string subjected to excessive pressure was found by survey to have failed in NINE places.[1] One failure is understandable. Two failures are imaginable, but NINE failures in one string of casing????? After a blowout in South Texas, the 9 5/8-inch surface casing was found to have parted at 3,200 feet and again at 1,600 feet. Combine conditions such as just described with the intense heat resulting from an oil-well fire or damage resulting from a falling derrick or collapsing substructure and it is easy to convince the average engineer that after a blowout the wellhead and tubulars could be expected to have little integrity. Properly applied, fluid dynamics can offer solutions which do not challenge the integrity of the tubulars in the blowout.

The applications of fluid dynamics to be considered are:

1. Kill-Fluid Bullheading
2. Kill-Fluid Lubrication
3. Dynamic Kill
4. Momentum Kill

KILL-FLUID BULLHEADING

"Bullheading" is the pumping of the kill fluid into the well against any pressure and regardless of any resistance the well may offer. Kill-fluid bullheading is one of the most common misapplications of fluid dynamics. Because bullheading challenges the integrity of the wellhead and tubulars, the result can cause further deterioration of the condition of the blowout. Many times wells have been lost, control delayed or options eliminated by the inappropriate bullheading of kill fluids.

Consider the following for an example of a proper application of the bullheading technique. During the development of the Ahwaz Field in Iran in the early 1970s, classic pressure control procedures were not possible. The producing horizon in the Ahwaz Field is so prolific that the difference between circulating and loosing circulation is a few psi. The typical wellbore schematic is presented as Figure 5.1. Drilling in the pay zone was possible by delicately balancing the hydrostatic with the formation pore pressure. The slightest underbalance resulted in a significant kick. Any classic attempt to control the well was unsuccessful because even the slightest back pressure at the surface caused lost circulation at the 9 5/8-inch casing shoe. Routinely, control was regained by increasing the weight of two hole volumes of mud at the surface by .1 to .2 ppg and pumping down the annulus to displace the influx and several hundred barrels of mud into the productive formation. Once the influx was displaced, routine drilling operations were resumed.

After the blowout at the Shell Cox in the Piney Woods of Mississippi, a similar procedure was adopted in the deep Smackover tests. In these operations, bringing the formation fluids to the surface was hazardous due to the high pressures and high concentrations of hydrogen sulfide. In response to the challenge, casing was set in the top of the Smackover. When a kick was taken, the influx was overdisplaced back into the Smackover by bullheading kill-weight mud down the annulus.

The common ingredients of success in these two examples are pressure, casing seat, and kick size. The surface pressures required to pump into the formation were low because the kick sizes were always small. In addition, it was of no consequence that the formation was fractured in the process and damaged by the mud pumped. The most important aspect was the casing seat. The casing that sat at the top of the productive interval in each example ensured that the kill mud as well as the influx would be forced back into the interval from which the kick occurred. It is most important to understand that, when bullheading, the kill fluid will almost always exit the wellbore at the casing seat.

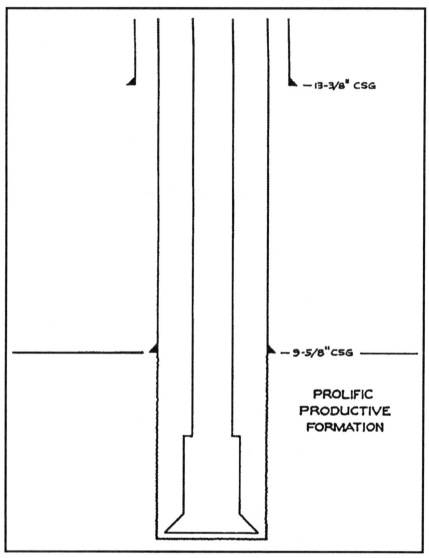

Figure 5.1 - Ahwaz Field

Consider an example of misapplication from the Middle East:

Example 5.1

  Given:

    Figure 5.2

Depth,	D	=	10,000 feet
Mud weight,	ρ	=	10 ppg
Mud gradient,	ρ_m	=	0.052 psi/ft
Gain		=	25 bbl
Hole size,	D_h	=	8 ½ inches
Intermediate casing (9 5/8-inch)		=	5,000 feet
Fracture gradient,	F_g	=	0.65 psi/ft
Fracture pressure:			
@ 5,000 ft,	$P_{frac5000}$	=	3250 psi
@ 10,000 ft,	$P_{frac10000}$	=	6500 psi
Shut-in annulus pressure,	P_a	=	400 psi
Drillpipe		=	4 ½ inch
Internal diameter,	D_i	=	3.826 inch
Drill collars:		=	800 feet
		=	6 inches x 2.25 inches
Gas specific gravity,	S_g	=	0.60
Temperature gradient		=	1.2°/100 ft

Ambient temperature		=	60 °F
Compressibility factor,	z	=	1.00

Capacity of:

Drill collar annulus,	C_{dcha}	=	0.0352 bbl/ft
Drillpipe,	C_{dpha}	=	0.0506 bbl/ft

The decision was made to weight up the mud to the kill weight and displace down the annulus since the drillpipe was plugged.

From Figure 5.2 and Equation 2.7:

$$P_b = \rho_f h + \rho_m (D - h) + P_a$$

and from Equation 3.7:

$$h_b = \frac{Influx\ Volume}{C_{dcha}}$$

$$h_b = \frac{25}{0.0352}$$

$$h_b = 710\ \textbf{feet}$$

and from Equation 3.5:

$$\rho_f = \frac{S_g P_b}{53.3 z_b T_b}$$

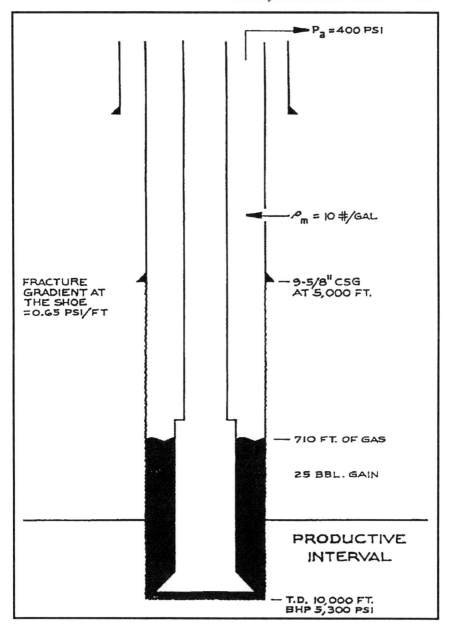

Figure 5.2

$$\rho_f = \frac{(0.6)(5200)}{53.3(1.0)(640)}$$

$\rho_f = 0.091$ psi/ft

Therefore:

$$P_b = 400 + 0.52(10000 - 710) + (0.091)(710)$$

$P_b = 5295$ psi

The kill-mud weight would then be given from Equation 2.9:

$$\rho_1 = \frac{P_b}{0.052D}$$

$$\rho_1 = \frac{5295}{(10000)(0.052)}$$

$\rho_1 = 10.2$ ppg

The annular capacity:

$$Capacity = (9200)(0.0506) + (800)(0.0352)$$

$Capacity = 494$ bbls

As a result of these calculations, 500 barrels of mud was weighted up at the surface to the kill weight of 10.2 ppg and pumped down the annulus. After pumping the 500 barrels of 10.2-ppg mud, the well was shut in and surface pressure was observed to be 500 psi, or 100 psi more than the 400 psi originally observed!

Required:

1. Explain the cause of the failure of the bullhead operation.

2. Explain the increase in the surface pressure.

Solution:

1.   The pressure at 5,000 feet is given by:

$$P_{5000} = (0.052)(10.0)(5000) + 400$$

$$P_{5000} = \textbf{3000 psi}$$

The fracture gradient at 5,000 feet is 3,250 psi. Therefore the difference is

$$\Delta P_{5000} = 3250\text{-}3000$$

$$\Delta P_{5000} = \textbf{250 psi}$$

The pressure at 10,000 feet is 5,296 psi. The fracture gradient at 10,000 feet is 6,500 psi. Therefore the difference is

$$\Delta P_{10000} = 6500\text{-}5296$$

$$\Delta P_{10000} = \textbf{1204 psi}$$

The pressure required to pump into the zone at the casing shoe was only 250 psi above the shut-in pressure and 954 psi less than the pressure required to pump into the zone at 10,000 feet. Therefore, when pumping operations commenced, the zone at the casing seat fractured and the mud was pumped into that zone.

2.   The surface pressure after the bullheading operation was 100 psi more than the surface pressure before the pumping job (500 psi versus 400 psi) because the 25 bbl influx had risen during the pumping operation. (See Chapter 4 on bubble rise for a further discussion of this topic.)

There are other reasons that a bullheading operation can fail. For example, after a well has been shut in, often the influx migrates to the surface, leaving drilling mud opposite the kick zone. Once pumping begins, the surface pressure must be increased until the zone to be bullheaded into is fractured by the drilling mud. The fracture pressure may be several hundred to several thousand pounds per square inch above the shut-in pressure. This additional pressure may be enough to rupture the casing in the well and cause an underground blowout.

Sometimes bullheading operations are unsuccessful when an annulus in a well is completely filled with gas that is to be pumped back into the formation. The reason for the failure in an instance such as this is that the kill mud bypasses the gas in the annulus during the pumping operation. Therefore, after the kill mud is pumped and the well is shut in to observe the surface pressure, there is pressure at the surface and gas throughout the system. The result is that the well unloads and blows out again.

Another consideration is the rate at which the mud being bullheaded is pumped. In the discussion concerning influx migration, it was noted that the influx most commonly migrates up one side of the annulus while the mud falls down the other side of the annulus. Further, when the influx nears the surface, the velocity of migration can be very high as evidenced by the rate of surface pressure increase. Under those conditions, bullheading at ¼ barrel per minute will not be successful because the mud will simply bypass the migrating influx. This is particularly problematic when the annulus area is large. The bullheading rate may have to be increased to more than 10 barrels per minute in order to be successful.   In any event, the bullheading rate will have to be increased until the shut-in surface pressure is observed to be decreasing as the mud is bullheaded into the annulus.

Bullheading is often used in deep, high-pressure well control situations to maintain acceptable surface pressures.   Consider Example 5.2 during underground blowouts:

Example 5.2
   Given:

Depth,	D	=	20,000 feet
Bottomhole pressure,	P_b	=	20000 psi
Casing shoe at,	D_{shoe}	=	10,000 feet
Fracture gradient at shoe,	F_g	=	0.9 psi/ft

The production is gas and the well is blowing out underground.

Required:
   Approximate the surface pressure if the gas is permitted to migrate to the surface.

   Determine the surface pressure if 15-ppg mud is continuously bullheaded into the annulus.

Solution:
   If the gas is permitted to migrate to the surface, the surface pressure will be approximately as follows:

$$P_{surf} = (F_g - 0.10)D_{shoe}$$

$$P_{surf} = (0.90 - 0.10)(10000)$$

$$P_{surf} = \textbf{8000 psi}$$

With 15-ppg mud bullheaded from the surface to the casing shoe at 10,000 feet, the surface pressure would be

$$P_{surf} = 8000 - (0.052)(15)(10000)$$

$$P_{surf} = 200 \text{ psi}$$

Therefore, as illustrated in Example 5.2, without the bullheading operation, the surface pressure would build to 8000 psi. At that pressure surface operations are very difficult at best. If 15-ppg mud is bullheaded into the lost circulation zone at the casing shoe, the surface pressure can be reduced to 200 psi. With 200 psi surface pressure, all operations such as snubbing or wire line are considerably easier and faster.

In summary, bullheading operations can have unpleasant results and should be thoroughly evaluated prior to commencing the operation. Too often, crews react to well control problems without analyzing the problem and get into worse condition than when the operation began. Remember, the best well control procedure is one that has predictable results from the technical as well as the mechanical perspective.

KILL-FLUID LUBRICATION — VOLUMETRIC KILL PROCEDURE

Kill-Fluid Lubrication, also sometimes called the Volumetric Kill Procedure, is the most overlooked well control technique. Lubricating the kill fluid into the wellbore involves an understanding of only the most fundamental aspects of physics. Basically, Kill-Fluid Lubrication is a technique whereby the influx is replaced by the kill fluid while the bottomhole pressure is maintained at or above the formation pressure. The result of the proper Kill-Fluid Lubrication operation is that the influx is removed from the wellbore, the bottomhole pressure is controlled by the kill-mud hydrostatic and no additional influx is permitted during the operation.

Kill-Fluid Lubrication has application in a wide variety of operations. The only requirement is that the influx has migrated to the surface or, as in some instances, the wellbore is completely void of drilling mud. Kill-Fluid Lubrication has application in instances when the pipe is on bottom, the pipe is part-way out of the hole or the pipe is completely out of the hole. The technique is applicable in floating drilling operations where the rig is often required, due to weather or some other emergency, to hang off, shut in and move off the hole. In each instance, the principles are fundamentally the same.

Consider the following from a recent well control problem at a deep, high-pressure operation in southeastern New Mexico. A kick was taken while on a routine trip at 14,080 feet. The pipe was out of the hole when the crew observed that the well was flowing. The crew ran 1,500 feet of drill string back into the hole. By that time the well was flowing too hard for the crew to continue the trip into the hole and the well was shut in. A sizable kick had been taken. Subsequently, the drillpipe was stripped into the hole in preparation for a conventional kill operation. However, the back-pressure valve, placed in the drill string 1,500 feet above the bit to enable the drillpipe to be stripped into the hole, had become plugged during the stripping operation. In addition, during the time the snubbing unit was being rigged up and the drillpipe was being stripped to bottom, the gas migrated to the surface. The influx came from a prolific interval at 13,913 feet. The zone had been drill-stem tested in this wellbore and had flowed gas at a rate of 10 mmscfpd with a flowing surface pressure of 5100 psi and a shut-in bottomhole pressure of 8442 psi. Since it was not possible to circulate the influx out of the wellbore in a classic manner, kill fluid was lubricated into the wellbore while efforts were being made to remove the obstruction in the drillpipe. The conditions as they existed at this location on that pleasant November afternoon are schematically illustrated in Figure 5.3. The following example illustrates the proper procedure for lubricating kill fluid into a wellbore:

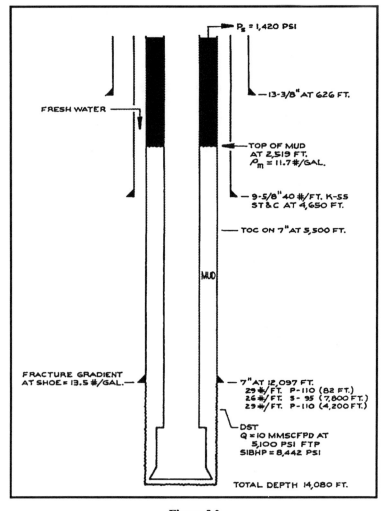

P_s = 1,420 PSI

— 13-3/8" AT 626 FT.

FRESH WATER

TOP OF MUD
AT 2,519 FT.
ρ_m = 11.7#/GAL.

— 9-5/8" 40 #/FT. K-55
ST&C AT 4,650 FT.

— TOC ON 7" AT 5,500 FT.

MUD

FRACTURE GRADIENT
AT SHOE = 13.5 #/GAL.—

— 7" AT 12,097 FT.
29 #/FT. P-110 (82 FT.)
26 #/FT. S- 95 (7,800 FT.)
29 #/FT. P-110 (4,200 FT.)

DST
Q = 10 MMSCFPD AT
5,100 PSI FTP
SIBHP = 8,442 PSI

TOTAL DEPTH 14,080 FT.

Figure 5.3

Example 5.3
  Given:
    Figure 5.3

Depth,	D	=	14,080 feet
Surface pressure,	P_a	=	1420 psi
Mud weight,	ρ	=	11.7 ppg

Fracture gradient at shoe, F_g = 0.702 psi/ft

Intermediate casing:

7-inch casing @, D_{shoe} = 12,097 feet

29 #/ft P-110   82 feet

26 #/ft S-95   7,800 feet

29 #/ft P-110   4,200 feet

Gas gravity, S_g = 0.6

Bottomhole pressure, P_b = 8442 psi

Temperature, T_s = 540 °Rankine

Kill-mud weight, ρ_1 = 12.8 ppg

Compressibility factor, z_s = 0.82

Capacity of:

Drillpipe annulus, C_{dpca} = 0.0264 bbl/ft

Drill-stem test at 13,913 feet

Volume rate of flow, Q = 10 mmscfpd @ 5100 psi

Plugged drillpipe at 12,513 feet

Required:

Design a procedure to lubricate kill mud into and the gas influx out of the annulus.

Solution:

Determine the height of the gas bubble, h, as follows from Equations 2.7 and 3.5:

$$P_b = \rho_f h + \rho_m (D-h) + P_a$$

$$\rho_f = \frac{S_g P_s}{53.3 z_s T_s}$$

$$\rho_f = \frac{(0.6)(1420)}{53.3(0.82)(540)}$$

$$\rho_f = 0.035 \text{ psi/ft}$$

Solve for, h, using Equation 2.7:

$$8442 = 1420 + (0.052)(11.7)(13913 - h) + 0.035h$$

$$h = 2{,}520 \text{ feet}$$

Gas volume at the surface, V_s:

$$V_s = (2520)(0.0264)$$

$$V_s = 66.5 \text{ bbls}$$

Determine the margin for pressure increase at the casing shoe using Equation 5.1:

$$P_{shoe} = \rho_f h + P_a + \rho_m (D_{shoe} - h) \tag{5.1}$$

Where:

P_{shoe}	= Pressure at the casing shoe, psi
ρ_f	= Influx gradient, psi/ft

h = Height of the influx, feet
P_a = Annulus pressure, psi
ρ_m = Original mud gradient, psi/ft
D_{shoe} = Depth to the casing shoe, feet

$$P_{shoe} = 0.036(2520) + 1420 + 0.6087(12097 - 2520)$$

$$P_{shoe} = \textbf{7340 psi}$$

Determine maximum permissible pressure at shoe, P_{frac}:

$$P_{frac} = F_g D_{shoe} \qquad\qquad (5.2)$$

$$P_{frac} = (0.052)(13.5)(12097)$$

$$P_{frac} = \textbf{8492 psi}$$

Where:
F_g = Fracture gradient, psi/ft
D_{shoe} = Depth of the casing shoe, feet

Maximum increase in surface pressure and hydrostatic, ΔP_t, that will not result in fracturing at the shoe is given by Equation 5.3:

$$\Delta P_t = P_{frac} - P_{shoe} \qquad\qquad (5.3)$$

$$\Delta P_t = 8492 - 7340$$

$$\Delta P_t = \textbf{1152 psi}$$

Where:

P_{frac} = Fracture pressure at casing shoe, psi
P_{shoe} = Calculated pressure at casing shoe, psi

The volume of the kill-weight mud, V_1, with density, ρ_1, to achieve, ΔP_t, is given by Equation 5.4:

$$V_1 = X_1 - \left[X_1 - \frac{\Delta P_t C_{dpca} V_s}{\rho_{m1}} \right]^{\frac{1}{2}} \tag{5.4}$$

$$X_1 = \frac{\rho_{m1} V_s + C_{dpca}(P_a - \Delta P_t)}{2(\rho_{m1})} \tag{5.5}$$

$$X_1 = \frac{0.667(66.5) + 0.0264(1420 + 1152)}{2(0.667)}$$

$$X_1 = 84.150$$

$$V_1 = 84.150 - \left[84.150^2 - \frac{1152(0.0264)(66.5)}{0.667} \right]^{\frac{1}{2}}$$

$$V_1 = 20.5 \text{ bbls}$$

Where:

ΔP_t = Maximum surface pressure, psi
C_{dpca} = Annular capacity, bbl/ft
V_s = Gas volume at the surface, bbl
ρ_{m1} = Kill mud gradient, psi/ft
X_1 = Intermediate calculation

Determine the effect of pumping 20 bbls of kill mud with density $\rho_1 = 12.8$ ppg. The resulting additional hydrostatic, ΔHyd, is calculated with Equation 5.6:

$$\Delta Hyd = 0.052 \, \rho_1 \left(\frac{V_1}{C_{dpca}} \right) \tag{5.6}$$

$$\Delta Hyd = (0.052)(12.8) \left(\frac{20}{0.0264} \right)$$

$$\Delta Hyd = \textbf{504 psi}$$

Additional surface pressure resulting from compressing the bubble at the surface with 20 bbls of kill mud is given by Equation 2.3:

$$\frac{P_1 V_1}{z_1 T_1} = \frac{P_2 V_2}{z_2 T_2}$$

1 - Prior to pumping kill mud

2 - After pumping kill mud

Therefore, by modifying Equation 2.2

$$P_2 = \frac{P_1 V_1}{V_2}$$

$$P_2 = \frac{(1420)(66.5)}{(66.5-20)}$$

$P_2 = $ **2031 psi**

Additional surface pressure, ΔP_s , is given as:

$$\Delta P_s = P_2 - P_a \tag{5.7}$$

$$\Delta P_s = 2031 - 1420$$

$\Delta P_s = $ **611 psi**

Total pressure increase, ΔP_{total}, is given as:

$$\Delta P_{total} = \Delta Hyd + \Delta P_s \tag{5.8}$$

$$\Delta P_{total} = 504 + 61$$

$\Delta P_{total} = $ **1115 psi**

Since ΔP_{total} is less than the maximum permissible pressure increase, ΔP_t, calculated using Equation 5.3, pump 20 bbls of 12.8-ppg mud at 1 bpm and shut in to permit the gas to migrate to the surface.

Observe initial P_2 after pumping = 1950 psi.

Observe 2-hour shut-in P_2 = 2031 psi.

The surface pressure, P_a, may now be reduced by bleeding **ONLY GAS** from P_2 to

$$P_{newa} = P_a - \Delta Hyd$$

$$P_{newa} = 1420 - 504$$

$$P_{newa} = \textbf{916 psi}$$

However, the new surface pressure must be used to determine the effective hydrostatic pressure at 13,913 feet to ensure no additional influx.

Equation 2.7 expands to Equation 5.9:

$$P_b = P_a + \rho_f h + \rho_m (D - h - h_1) + \rho_{m1} h_1 \qquad (5.9)$$

$$h = \frac{V_s}{C_{dpca}}$$

Where:

$\qquad V_s \qquad$ = Remaining volume of influx, bbls

$$h = \frac{66.5 - 20}{0.0264}$$

$$h = \textbf{1,762 feet}$$

$$h_1 = \frac{V_1}{C_{dpca}}$$

$$h_1 = \frac{20}{0.0264}$$

h_1 = **758 feet**

$$\rho_f = \frac{S_g P_s}{53.3 z_s T_s}$$

$$\rho_f = \frac{(0.6)(916)}{(53.3)(0.866)(540)}$$

ρ_f = **0.022 psi/ft**

$$P_b = 916 + (0.6084)(13913 - 1762 - 758)$$
$$+ (0.022)(1762) + (0.667)(758)$$

P_b = **8392 psi**

However, the shut-in pressure is 8442 psi. Therefore, since the effective hydrostatic cannot be less than the reservoir pressure, the surface pressure can only be bled to

$$P_a = 916 + (8442 - 8392)$$

P_a = **966 psi**

Important! The pressure cannot be reduced by bleeding mud. If mud is bled from the annulus, the well must be shut in for a longer period to allow the gas to migrate to the surface.

Now, the procedure must be repeated until the influx is lubricated from the annulus and replaced by mud:

$$\rho_f = \frac{S_g P_s}{53.3 z_s T_s}$$

$$\rho_f = \frac{(0.6)(966)}{53.3(0.866)(540)}$$

$$\rho_f = \textbf{0.023 psi/ft}$$

Now solving for h, use

$$h = \frac{V_s}{C_{dpca}}$$

$$h = \textbf{1,762 feet}$$

Similarly solving for h_1:

$$h_1 = \frac{V_1}{C_{dpa}}$$

$$h_1 = \frac{20}{0.0264}$$

$$h_1 = \textbf{758 feet}$$

Determine the margin for pressure increase at the casing shoe using Equation 5.10, which is Equation 5.1 modified for inclusion of kill mud, ρ_1:

$$P_{shoe} = \rho_f h + P_a + \rho_m (D_{shoe} - h - h_1) + \rho_1 h_1 \qquad (5.10)$$

$$P_{shoe} = 0.023(1762) + 966 + 0.6087$$
$$(12097 - 1762 - 758) + (0.052)(12.8)(758)$$

$$P_{shoe} = 7339 \text{ psi}$$

The maximum permissible pressure at shoe, P_{frac}, from Equation 5.2 is equal to 8492 psi.

The maximum increase in surface pressure and hydrostatic, ΔP_t, that will not result in fracturing at the shoe is given by Equation 5.3:

$$\Delta P_t = P_{frac} - P_{shoe}$$

$$\Delta P_t = 8492 - 7339$$

$$\Delta P_t = 1153 \text{ psi}$$

The volume of the kill-weight mud, V_1, with density, ρ_1, to achieve ΔP_t is given by Equations 5.4 and 5.5:

$$V_1 = X_1 - \left[X_1 - \frac{\Delta P_t C_{dpca} V_s}{\rho_{m1}} \right]^{\frac{1}{2}}$$

$$X_1 = \frac{\rho_{ml} V_s + C_{dpca}(P_a - \Delta P_t)}{2(\rho_{ml})}$$

$$X_1 = \frac{0.667(46.5) + 0.0264(966 + 1153)}{2(0.667)}$$

$$X_1 = 65.185$$

$$V_1 = 65.185 - \left[65.185^2 - \frac{1153(0.0264)(46.5)}{0.667} \right]^{\frac{1}{2}}$$

$$V_1 = 19.1 \text{ bbls}$$

Determine the effect of pumping 19 bbls of kill mud with density $\rho_1 = 12.8$ ppg. The resulting additional hydrostatic, ΔHyd, is

$$\Delta Hyd = (0.052)(\rho_1)\left(\frac{V_1}{C_{dpca}}\right)$$

$$\Delta Hyd = (0.052)(12.8)\left(\frac{19}{0.0264}\right)$$

$$\Delta Hyd = 480 \text{ psi}$$

Additional surface pressure resulting from compressing the bubble at the surface with 19 bbls of kill mud is

$$P_2 = \frac{P_1 V_1}{V_2}$$

$$P_2 = \frac{(966)(46.5)}{(46.5 - 19)}$$

$$P_2 = 1633 \text{ psi}$$

Additional surface pressure, ΔP_s, is given as

$$\Delta P_s = P_2 - P_a$$

$$\Delta P_s = 1633 - 966$$

$$\Delta P_s = 667 \text{ psi}$$

Total pressure increase, ΔP_{total}, is given by Equation 5.8:

$$\Delta P_{total} = \Delta Hyd + \Delta P_s$$

$$\Delta P_{total} = 480 + 667$$

$$\Delta P_{total} = 1147 \text{ psi}$$

Since ΔP_{total} is less than the maximum permissible pressure increase, ΔP_t, calculated using Equation 5.3, pump 19 bbls of 12.8-ppg mud at 1 bpm and shut in to permit the gas to migrate to the surface.

Observe initial P_2 after pumping = 1550 psi.

Observe 2-hour shut-in $P_2 = 1633$ psi.

The surface pressure, P_a, may now be reduced by bleeding **ONLY GAS** from P_2 to

$$P_{newa} = P_a - \Delta Hyd$$

$$P_{newa} = 966 - 480$$

$$P_{newa} = \textbf{486 psi}$$

However, the new surface pressure must be used to determine the effective hydrostatic pressure at 13,913 feet to ensure no additional influx.

The bottomhole pressure is given by Equation 5.9:

$$P_b = P_a + \rho_f h + \rho_m (D - h - h_1) + \rho_{m1} h_1$$

$$h = \frac{V_s}{C_{dpca}}$$

$$h = \frac{46.5 - 19}{0.0264}$$

$$h = \textbf{1,042 feet}$$

$$h_1 = \frac{V_1}{C_{dpca}}$$

$$h_1 = \frac{39}{0.0264}$$

$$h_1 = 1{,}477 \text{ feet}$$

$$\rho_f = \frac{S_g P_s}{53.3 z_s T_s}$$

$$\rho_f = \frac{(0.6)(486)}{(53.3)(0.930)(540)}$$

$$\rho_f = 0.011 \text{ psi/ft}$$

$$P_b = 486 + (0.6084)(13913 - 1042 - 1477) \\ + (0.011)(1042) + (0.667)(1477)$$

$$P_b = 8415 \text{ psi}$$

However, the shut-in pressure is 8442 psi. Therefore, since the effective hydrostatic cannot be less than the reservoir pressure, the surface pressure can only be bled to

$$P_a = 486 + (8442 - 8415)$$

$$P_a = 513 \text{ psi}$$

The surface pressure must be bled to only 513 psi to ensure no additional influx by bleeding only dry gas through the choke manifold. **Important! No mud can be bled from the annulus.**

If the well begins to flow mud from the annulus, it must be shut in until the gas and mud separate.

Again, the procedure must be repeated:

$$\rho_f = \frac{S_g P_s}{53.3 z_s T_s}$$

$$\rho_f = \frac{(0.6)(513)}{53.3(0.926)(540)}$$

$$\rho_f = 0.012 \text{ psi/ft}$$

Now solving for h, use

$$h = \frac{V_s}{C_{dpca}}$$

$$h = \frac{46.5 - 19.0}{0.0264}$$

$$h = 1,042 \text{ feet}$$

Similarly solving for h_1:

$$h_1 = \frac{V_1}{C_{dpca}}$$

$$h_1 = \frac{39}{0.0264}$$

$$h_1 = 1,477 \text{ feet}$$

Determine the margin for pressure increase at the casing shoe using Equation 5.10:

$$P_{shoe} = \rho_f h + P_a + \rho_m (D_{shoe} - h - h_1) + \rho_1 h_1$$

$$P_{shoe} = 0.012(1042) + 513 + 0.6084$$
$$(12097 - 1042 - 1477) + (0.052)(12.8)(1477)$$

$$P_{shoe} = 7336 \text{ psi}$$

The maximum permissible pressure at shoe, P_{frac}, from Equation 5.2 is equal to 8492 psi.

The maximum increase in surface pressure and hydrostatic, ΔP_t, that will not result in fracturing at the shoe is given by Equation 5.3:

$$\Delta P_t = P_{frac} - P_{shoe}$$

$$\Delta P_t = 8492 - 7336$$

$$\Delta P_t = 1156 \text{ psi}$$

The volume of the kill-weight mud, V_1, with density, ρ_1, to achieve, ΔP_t, is given by Equations 5.4 and 5.5:

$$V_1 = X_1 - \left[X_1 - \frac{\Delta P_t C_{dpca} V_s}{\rho_{ml}} \right]^{\frac{1}{2}}$$

$$X_1 = \frac{\rho_{ml} V_s + C_{dpca}(P_a - \Delta P_t)}{2(\rho_{ml})}$$

$$X_1 = \frac{0.667(27.5) + 0.0264(513 + 1156)}{2(0.667)}$$

$$X_1 = 46.780$$

$$V_1 = 46.780 - \left[46.780^2 - \frac{1156(0.0264)(27.5)}{0.667} \right]^{\frac{1}{2}}$$

$$V_1 = 16.3 \text{ bbls}$$

Determine the effect of pumping 16 bbls of kill mud with density $\rho_1 = 12.8$ ppg. The resulting additional hydrostatic, ΔHyd, is given by Equation 5.6:

$$\Delta Hyd = 0.052 \rho_1 \left(\frac{V_1}{C_{dpca}} \right)$$

$$\Delta Hyd = (0.052)(12.8)\left(\frac{16}{0.0264} \right)$$

$$\Delta Hyd = 404 \text{ psi}$$

Additional surface pressure resulting from compressing the bubble at the surface with 16 bbls of kill mud is

$$P_2 = \frac{P_1 V_1}{V_2}$$

$$P_2 = \frac{(513)(27.5)}{(27.5 - 16)}$$

$$P_2 = 1227 \text{ psi}$$

Additional surface pressure, ΔP_s is given by Equation 5.7:

$$\Delta P_s = P_2 - P_a$$

$$\Delta P_s = 1227 - 513$$

$$\Delta P_s = 714 \text{ psi}$$

Total pressure increase, ΔP_{total}, is given by Equation 5.8:

$$\Delta P_{total} = \Delta Hyd + \Delta P_s$$

$$\Delta P_{total} = 404 + 714$$

$$\Delta P_{total} = 1118 \text{ psi}$$

Since ΔP_{total} is less than the maximum permissible pressure increase, ΔP_t, calculated using Equation 5.3, pump 16 bbls of 12.8-ppg mud at 1 bpm and shut in to permit the gas to migrate to the surface.

In 2 to 4 hours shut-in $P_2 = 1227$ psi.

The surface pressure, P_a, may now be reduced by bleeding **ONLY GAS** from P_2 to

$$P_{newa} = P_a - \Delta Hyd$$

$$P_{newa} = 513 - 404$$

$$P_{newa} = \textbf{109 psi}$$

However, the new surface pressure must be used to determine the effective hydrostatic pressure at 13,913 feet to insure no additional influx.

Pursuant to Equation 5.9:

$$P_b = P_a + \rho_f h + \rho_m (D - h - h_1) + \rho_{m1} h_1$$

$$h = \frac{V_s}{C_{dpca}}$$

$$h = \frac{27.5 - 16}{0.0264}$$

$$h = \textbf{436 feet}$$

$$h_1 = \frac{V_1}{C_{dpca}}$$

$$h_1 = \frac{55}{0.0264}$$

$$h_1 = 2,083 \text{ feet}$$

$$\rho_f = \frac{S_g P_s}{53.3 z_s T_s}$$

$$\rho_f = \frac{(0.6)(109)}{(53.3)(0.984)(540)}$$

$$\rho_f = 0.0023 \text{ psi/ft}$$

$$P_b = 109 + (0.6084)(13913 - 436 - 2083)$$
$$+ (0.011)(436) + (0.667)(2083)$$

$$P_b = 8431 \text{ psi}$$

However, the shut-in pressure is 8442 psi. Therefore, since the effective hydrostatic cannot be less than the reservoir pressure, the surface pressure can only be bled to

$$P_a = 109 + (8442 - 8431)$$

$$P_a = 120 \text{ psi}$$

The surface pressure must be bled to only 120 psi to ensure no additional influx by bleeding only dry gas through the choke manifold. Important! No mud can be bled from the annulus.

If the well begins to flow mud from the annulus, it must be shut in until the gas and mud separate.

Now the procedure is repeated for the final increment:

$$\rho_f = \frac{S_g P_s}{53.3 z_s T_s}$$

$$\rho_f = \frac{(0.6)(120)}{53.3(0.983)(540)}$$

$$\rho_f = 0.0025 \text{ psi/ft}$$

Now solving for h, use

$$h = \frac{V_s}{C_{dpca}}$$

$$h = \frac{27.5 - 16.0}{0.0264}$$

$$h = 436 \text{ feet}$$

Similarly solving for h_1:

$$h_1 = \frac{V_1}{C_{dpca}}$$

$$h_1 = \frac{55}{0.0264}$$

$$h_1 = 2,083 \text{ feet}$$

Determine the margin for pressure increase at the casing shoe using Equation 5.10:

$$P_{shoe} = \rho_f h + P_a + \rho_m (D_{shoe} - h - h_1) + \rho_1 h_1$$

$$P_{shoe} = 0.0025(436) + 120 + 0.6084$$
$$(12097 - 436 - 2083) + (0.052)(12.8)(2083)$$

$$P_{shoe} = 7338 \text{ psi}$$

The maximum permissible pressure at shoe, P_{frac}, from Equation 5.2 is equal to 8489 psi.

The maximum increase in surface pressure and hydrostatic, ΔP_t, that will not result in fracturing at the shoe is given by Equation 5.3:

$$\Delta P_t = P_{frac} - P_{shoe}$$

$$\Delta P_t = 8492 - 7338$$

$$\Delta P_t = 1154 \text{ psi}$$

The volume of the kill-weight mud, V_1, with density, ρ_1, to achieve, ΔP_t, is given by Equations 5.4 and 5.5:

$$V_1 = X_1 - \left[X_1 - \frac{\Delta P_t C_{dpca} V_s}{\rho_{m1}} \right]^{\frac{1}{2}}$$

$$X_1 = \frac{\rho_{ml} V_s + C_{dpca}(P_a - \Delta P_t)}{2(\rho_{ml})}$$

$$X_1 = \frac{0.667(11.5) + 0.0264(120 + 1154)}{2(0.667)}$$

$$X_1 = 30.963$$

$$V_1 = 30.963 - \left[30.963^2 - \frac{1154(0.0264)(11.5)}{0.667} \right]^{\frac{1}{2}}$$

$$V_1 = 10.14 \text{ bbls}$$

Determine the effect of pumping 10 bbls of kill mud with density $\rho_1 = 12.8$ ppg. The resulting additional hydrostatic, ΔHyd, is given by Equation 5.6:

$$\Delta Hyd = 0.052 \, \rho_1 \left(\frac{V_1}{C_{dpca}} \right)$$

$$\Delta Hyd = (0.052)(12.8) \left(\frac{10}{0.0264} \right)$$

$$\Delta Hyd = 252 \text{ psi}$$

Additional surface pressure resulting from compressing the bubble at the surface with 10 bbls of kill mud is

$$P_2 = \frac{P_1 V_1}{V_2}$$

$$P_2 = \frac{(120)(11.5)}{(11.5-10)}$$

$$P_2 = \textbf{920 psi}$$

Additional surface pressure, ΔP_s is given as

$$\Delta P_s = P_2 - P_a$$

$$\Delta P_s = 920 - 120$$

$$\Delta P_s = \textbf{800 psi}$$

Total pressure increase, ΔP_{total}, is given as

$$\Delta P_{total} = \Delta Hyd + \Delta P_s$$

$$\Delta P_{total} = 252 + 800$$

$$\Delta P_{total} = \textbf{1052 psi}$$

Since ΔP_{total} is less than the maximum permissible pressure increase, ΔP_t, calculated using Equation 5.3, pump 10 bbls of 12.8-ppg mud at 1 bpm and shut in to permit the gas to migrate to the surface.

In 2 to 4 hours shut-in $P_2 = 920$ psi.

The surface pressure, P_a, may now be reduced by bleeding **ONLY GAS** from P_2 to

$$P_{newa} = P_a - \Delta Hyd$$

$$P_{newa} = 120 - 252$$

$$P_{newa} = 0 \text{ psi}$$

and the hole filled and observed.

The well is killed.

Summary

	Volume Pumped	Old Surface Pressure	Surface Pressure after Adding Kill Mud	Surface Pressure after Bleeding
Initial Conditions	-0-	1420		
1st Stage	20	1420	2031	966
2nd Stage	19	966	1633	513
3rd Stage	16	513	1227	120
4th Stage	10	120	920	-0-

By following this schedule, the well can be killed safely without violating the casing seat and losing more mud and without permitting any additional influx of formation gas into the wellbore.

In summary, the well is controlled by pumping kill-weight mud into the annulus and bleeding dry gas out of the annulus. The volume of kill fluid that can be pumped without fracturing the casing seat is determined from Equation 5.3. Logically, the surface pressure should be reduced from the value prior to lubricating kill mud by the additional hydrostatic contributed by the kill-weight mud. However, for obvious reasons, the effective hydrostatic cannot be less than the reservoir pressure. At the lower surface pressures, the gas hydrostatic gradient is less and the final surface pressure must be higher to reflect the difference in gas gradient and prevent additional influx. The analysis is continued until the well is dead. Rudimentary analysis suggests higher mud weights for well control.

DYNAMIC KILL OPERATIONS

Simply put, to kill a well dynamically is to use frictional pressure losses to control the flowing bottomhole pressure and, ultimately, the static bottomhole pressure of the blowout.

Dynamic Kill implies the use of a kill fluid whose density results in a hydrostatic column which is less than the static reservoir pressure. Therefore, the frictional pressure loss of the kill fluid is required to stop the flow of reservoir fluids. Dynamic Kill in the purest sense was intended as an intermediate step in the well control procedure. Procedurally, after the blowout was dynamically controlled with a fluid of lesser density, it was ultimately controlled with a fluid of greater density, which resulted in a hydrostatic greater than the reservoir pressure. Generically and in this text, Dynamic Kill includes control procedures utilizing fluids with densities less than and much greater than that required to balance the static reservoir pressure. In reality, when the density of the kill fluid is equal to or greater than that required to balance the static reservoir pressure, the fluid dynamics are more properly described as a Multiphase Kill Procedure.

Dynamically controlling a well using a Multiphase Kill Procedure is one of the oldest and most widely used fluid control operations. In the past, it has been a "seat-of-the-pants" operation with little or no technical evaluation. Most of the time, well control specialists had some arbitrary rules of thumb, e.g. the kill fluid had to be 2 pounds per gallon heavier than the mud used to drill the zone or the kill fluid had to achieve some particular annular velocity. Usually, that all translated into manifolding together all the pumps in captivity, weighting the mud up as high as possible, pumping like hell and hoping for the best. Sometimes it worked and sometimes it didn't.

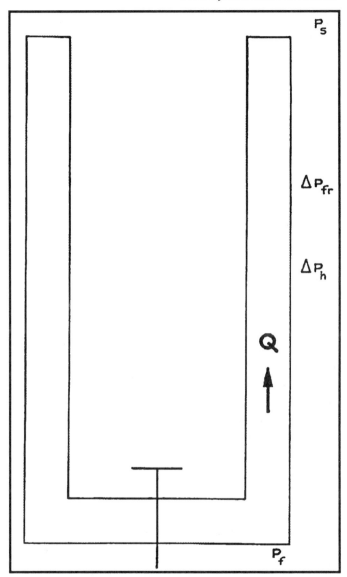

Figure 5.4 - Dynamic Kill Schematic

There are many applications of the Dynamic and Multiphase Kill procedures and any well control operation should be studied from that perspective. The most common application is when the well is flowing out of control and the drillpipe is on bottom. The kill fluid is pumped to the bottom of the hole through the drillpipe, and the additional hydrostatic of the kill fluid along with the increased friction pressure resulting from the kill fluid controls the well.

Multiphase Kill operations were routine in the Arkoma Basin in the early 1960s.   There, air drilling was popular.   In air drilling operations, every productive well is a blowout.   Under different circumstances, some of them would have made the front page of the *New York Times*.   At one location just east of McCurtain, Oklahoma, the fire was coming out of an 8-inch blooie line and was almost as high as the crown of the 141-foot derrick.   The blooie line was 300 feet long, and you could toast marshmallows on the rig floor.   The usual procedure was the seat-of-the-pants Multiphase Kill and it killed the well.

The most definitive work on the pure Dynamic Kill was done by Mobil Oil Corporation and reported by Elmo Blount and Edy Soeiinah.[2] The biggest gas field in the world is Mobil's Arun Field in North Sumatra, Indonesia.   On 4 June, 1978, Well No. C-II-2 blew out while drilling and caught fire.   The rig was immediately consumed.   The well burned for 89 days at an approximate rate of 400 million standard cubic feet per day.   Due to the well's high deliverability and potential, it was expected to be extremely difficult to kill.   The engineering was so precise that only one relief well was required.   That a blowout of this magnitude was completely dead one hour and 50 minutes after pumping operations commenced is a tribute to all involved.   One of the most significant contributions resulting from this job was the insight into the fluid dynamics of a Dynamic Kill.

The engineering concepts of a Dynamic Kill are best understood by considering the familiar U-tube of Figure 5.4. The left side of the U-tube may represent a relief well and the right side a blowout, as was the case just discussed, or the left side may represent drillpipe while the right side would correspond to the annulus, as is the case in many well control situations. The connecting interval may be the formation in the case of a relief well with the valve representing the resistance due to the flow of fluids through the formation.   In the case of the drillpipe - annulus scenario, the valve may represent the friction in the drill string or the nozzles in the bit, or the situation may be completely different. Whatever the situation, the technical concepts are basically the same.

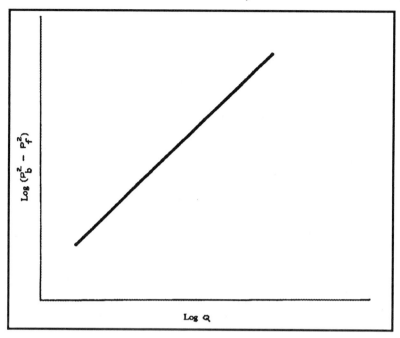

Figure 5.5 - *Typical Open Flow Potential Curve*

The formation is represented as flowing up the right side of the U-tube with a flowing bottomhole pressure, P_{flow}, which is given by the following equation:

$$P_{flow} = P_a + \Delta P_{fr} + \Delta P_h \qquad (5.11)$$

Where:

P_a = Surface pressure, psi

ΔP_{fr} = Frictional pressure, psi

ΔP_h = Hydrostatic pressure, psi

The capability of the formation to deliver hydrocarbons to the wellbore is governed by the familiar back-pressure curve illustrated in Figure 5.5 and described by the Equation 5.12:

$$Q = C\left(P_b^2 - P_{flow}^2\right)^n \qquad (5.12)$$

Where:

Q = Flow rate, mmscfpd
P_b = Formation pore pressure, psi
P_{flow} = Flowing bottomhole pressure, psi
C = Constant
n = Slope of back-pressure curve
      = 0.5 for turbulence
      = 1.0 for laminar flow

Finally, the reaction of the formation to an increase in flowing bottomhole pressure, P_{flow}, is depicted by the classic Horner Plot, illustrated as Figure 5.6.   The problem is to model the blowout considering these variables.

In the past, shortcuts have been taken for the sake of simplicity. For example, the most simple approach is to design a kill fluid and rate such that the frictional pressure loss plus the hydrostatic is greater than the shut-in bottomhole pressure, P_b. This rate would be that which is sufficient to maintain control. Equations 4.13 through 4.14 for frictional pressure losses in turbulent flow can be used in such an analysis.

Consider the Example 5.4:

Example 5.4
   Given:

Depth,	D	=	10,000 feet
Kill fluid,	ρ_1	=	8.33 ppg
Kill-fluid gradient	ρ_{ml}	=	0.433 psi/ft
Bottomhole pressure,	P_b	=	5200 psi
Inside diameter of pipe,	D_i	=	4.408 inches
Surface pressure,	P_a	=	atmospheric

   Figure 5.4

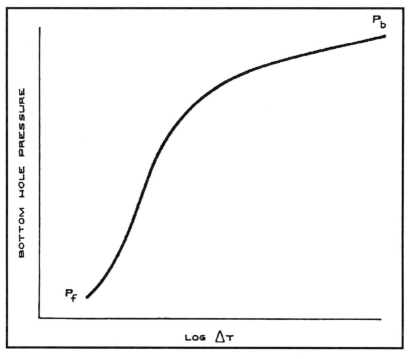

Figure 5.6 - *Typical Pressure Build Up Curve*

Required:

Determine the rate required to kill the well dynamically with water.

Solution:

$$P_{flow} = P_a + \Delta P_{fr} + \Delta P_h$$

$$\Delta P_h = 0.433(10000)$$

$$\Delta P_h = \textbf{4330 psi}$$

$$\Delta P_{fr} = 5200\text{-}4330$$

$$\Delta P_{fr} = \textbf{870 psi}$$

Rearranging Equation 4.14 where $\Delta P_{fr} = P_{fti}$:

$$Q = \left[\frac{P_{fti} D_i^{4.8}}{7.7(10^{-5}) \rho_1^{.8} PV^{.2} l} \right]^{\frac{1}{1.8}}$$

$$Q = \left[\frac{(870)(4.408)^{4.8}}{7.7(10^{-5})(8.33)^{.8}(1)^2(10000)} \right]^{\frac{1}{1.8}}$$

$$Q = 1{,}011 \text{ gpm}$$

$$Q = 24.1 \text{ bpm}$$

Therefore, as illustrated in Example 5.4, the well would be dynamically controlled by pumping fresh water through the pipe at 24 bpm. The Dynamic Kill Procedure would be complete when the water was followed with kill mud of sufficient density (10 ppg in this example) to control the bottomhole pressure.

The rate required to maintain control is insufficient in most instances to achieve control and is considered to be the minimum rate for a Dynamic Kill operation. Kouba, et. al. have suggested that the rate sufficient to maintain control is the minimum and that the maximum rate is approximated by Equation 5.13:[3]

$$Q_{k\,max} = A \left(\frac{2 g_c D_{tvd} D_h}{f D_{md}} \right)^{\frac{1}{2}} \tag{5.13}$$

Where:

g_c = Gravitational constant, $\dfrac{lb_m \cdot ft}{lb_f \cdot sec^2}$

A = Cross sectional area, ft^2

D_h = Hydraulic diameter, feet

D_{tvd} = Vertical well height, feet

$$D_{md} \quad = \quad \text{Measured well length, feet}$$
$$f \quad = \quad \text{Moody friction factor, dimensionless}$$

Consider Example 5.5:

Example 5.5

Given:

Same conditions as Example 5.4

Required:

Calculate the maximum kill rate using Equation 5.13.

Solution:

Solving Equation 5.13 gives

$$Q_{k\,max} = A\left(\frac{2g_c D_{tvd} D_h}{f D_{md}}\right)^{\frac{1}{2}}$$

$$Q_{k\,max} = 0.106\left(\frac{2(32.2)(10000)(.3673)}{.019(10000)}\right)^{\frac{1}{2}}$$

$$Q_{k\,max} = 3.74\ \frac{ft^3}{\text{sec}}$$

$$Q_{k\,max} = \textbf{40 bpm}$$

As illustrated in Examples 5.4 and 5.5, the minimum kill rate is 24 bpm and the maximum kill rate is 40 bpm. The rate required to kill the well is somewhere between these values and is very difficult to determine. Multiphase flow analysis is required. The methods of multiphase flow analysis are very complex and based upon empirical correlations obtained from laboratory research. The available correlations and research are based upon gas-lift models describing the flow of gas, oil and water inside small pipes. Precious little research has been intended to describe annular flow much less the multiphase relationship between gas, oil, drilling mud and water flowing up a very large, inclined annulus. The conditions and boundaries describing most blowouts are very complex to be described by currently available multiphase models. It is beyond the scope of this work to offer an in-depth discussion of multiphase models.

Further complicating the problem is the fact that, in most instances, the productive interval does not react instantaneously as would be implied by the strict interpretation of Figure 5.5. Actual reservoir response is illustrated by the classical Horner Plot illustrated in Figure 5.6. As illustrated in Figure 5.6, the response by the reservoir to the introduction of a kill fluid is non-linear. For example, the multiphase frictional pressure loss (represented by Figure 5.6) initially required to control the well is not that which will control the static reservoir pressure. The multiphase frictional pressure loss required to control the well is that which will control the flowing bottomhole pressure. The flowing bottomhole pressure may be much less than the static bottomhole pressure. Further, several minutes to several hours may be required for the reservoir to stabilize at the reservoir pressure. Unfortunately, much of the data needed to understand completely the productive capabilities of the reservoir in a particular wellbore are not available until after the blowout is controlled. However, data from similar offset wells can be considered.

Consider the well control operation at the Williford Energy Company Rainwater No. 2-14 in Pope County near Russellville, Arkansas.[4] The wellbore schematic is presented as Figure 5.7. A high-volume gas zone had been penetrated at 4,620 feet. On the trip out of the hole, the well kicked. Mechanical problems prevented the well from being shut in and it was soon flowing in excess of 20 mmscfpd through the rotary table. The drillpipe was stripped to bottom and the well was diverted through the choke manifold. By pitot tube, the well was determined to be flowing at a rate of 34.9 mmscfpd with a manifold pressure of 150 psig. The wellbore schematic, Open Flow Potential Test and Horner Plot are presented as Figures 5.7, 5.8 and 5.9, respectively.

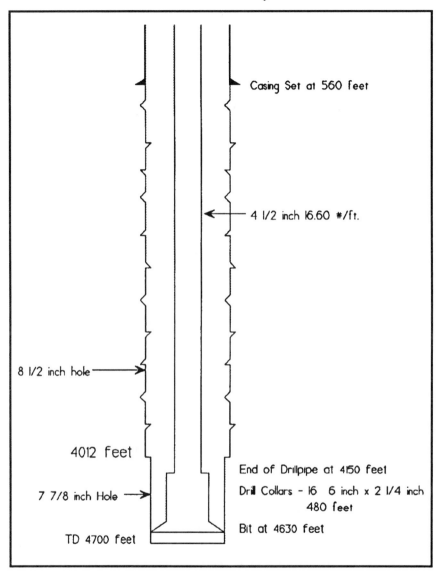

Casing Set at 560 feet

4 1/2 inch 16.60 #/ft.

8 1/2 inch hole

4012 feet

7 7/8 inch Hole

TD 4700 feet

End of Drillpipe at 4150 feet

Drill Collars - 16 6 inch x 2 1/4 inch
480 feet

Bit at 4630 feet

Figure 5.7 *- Williford Energy Company*
Rainwater No. 1, Pope County, Arkansas

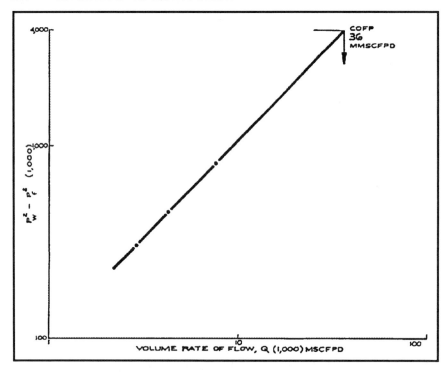

Figure 5.8 *- Calculated Open Flow Potential Test*
Williford Energy Company - Rainwater No. 2-14

In this instance, the Orkiszewski method was modified and utilized to predict the multiphase behavior.[5]   The well was successfully controlled.

The technique used in the Williford Energy example is very conservative in that it determines the multiphase kill rate required to control the shut in bottomhole pressure of 1995 psi. Analysis of Figure 5.9 indicates that the static bottomhole pressure will not be reached in the blowout for more than 100 hours. The flowing bottomhole pressure when the kill procedure begins is only 434 psi and is only 1500 psi approximately 20 minutes after the flowing bottomhole pressure has been exceeded. Of course, in this case the kill operation is finished in just over 10 minutes. Including all these variables is more complex but well within the capabilities of modern computing technology. Based on this more complex analysis, the kill rate using 10.7-ppg mud was determined to be 10 bpm.

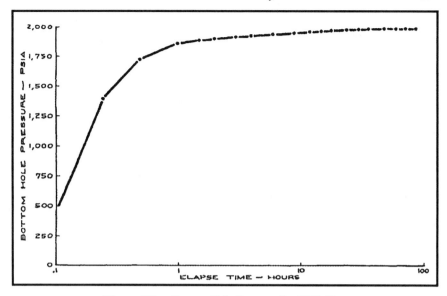

Figure 5.9 - *Bottom Hole Pressure Build Up Test*
Williford Energy Company - Rainwater No. 2-14

The model is further complicated by the fact that the Calculated Open Flow Potential curve (COFP) presented as Figure 5.8 is a very optimistic evaluation of sustained productive capacity. An Actual Open Flow Potential curve (AOFP) based on sustained production would be more appropriate for modeling actual kill requirements. A kill procedure based on the COFP is the more conservative approach. An AOFP curve more accurately reflects the effect of the pressure draw down in the reservoir in the vicinity of the wellbore.

THE MOMENTUM KILL

The Momentum Kill is a procedure where two fluids collide and the one with the greater momentum wins. If the greater momentum belongs to the fluid from the blowout, the blowout continues. If the greater momentum belongs to the kill fluid, the well is controlled. The technology of the Momentum Kill procedure is the newest and least understood of well control procedures. However, the technique itself is not new. In the late fifties and early sixties, the air drillers of eastern Oklahoma thought nothing of pulling into the surface pipe to mud up an air-drilled hole in an effort to avoid the hazards associated with the introduction of mud to the Atoka Shale.

Momentum Kill concepts are best illustrated by Figures 5.10 and 5.11. Figure 5.10 illustrates a situation in which the outcome would never be in doubt. The most fundamental reasoning would conclude that the occupant of the car is in greater peril than the occupant of the truck. It is most likely that the momentum of the truck will prevail and the direction of the car will be reversed. Conceptually, fluid dynamics are not as easy. However, consider the guys in Figure 5.11. They have forgotten the fire and turned their attention to each other. Obviously, the one with the least momentum is destined for a bath.

Figure 5.10

The dynamics of a blowout are very much the same as those illustrated in Figure 5.11. The fluid flowing from the blowout exhibits a definable quantity of momentum. Therefore, if the kill fluid is introduced at a greater momentum, the flow from the blowout is reversed when the fluids collide. The governing physical principles are not significantly different from those governing the collision of two trains, two cars or two men. The mass with the greatest momentum will win the encounter.

Figure 5.11

Newton's Second Law states that the net force acting on a given mass is proportional to the time rate of change of linear momentum of that mass. In other words, the net external force acting on the fluid within a prescribed control volume equals the time rate of change of momentum of the fluid within the control volume plus the net rate of momentum transport out of the surfaces of the control volume.

Consider the following development with all units being basic:

Momentum:

$$M = \frac{mv}{g_c} \tag{5.14}$$

and the mass rate of flow:

$$\omega = \rho v A = \rho q \tag{5.15}$$

and, from the conservation of mass:

$$\rho v A = \rho_i v_i A_i \tag{5.16}$$

Where:

m = Mass, lb_m

v = Velocity, $\dfrac{ft}{sec}$

g_c = Gravitational constant, $\dfrac{lb_m \cdot ft}{lb_f \cdot sec^2}$

ρ = Density, $\dfrac{lb_m}{ft^3}$

q = Volume rate of flow, $\dfrac{ft^3}{sec}$

ω = Mass rate of flow, $\dfrac{lb_m}{sec}$

i = Conditions at any point

A = Cross sectional flow area, ft^2

All variables are at standard conditions unless noted with a subscript, i.

The momentum of the kill fluid is easy to compute because it is essentially an incompressible liquid. The momentum of the kill fluid is given by Equation 5.14:

$$M = \frac{mv}{g_c}$$

Substituting

$$v = \frac{q}{A}$$

$$m = \omega = \rho q$$

results in the momentum of the kill fluid, Equation 5.17:

$$M = \frac{\rho q^2}{g_c A}$$ (5.17)

Since the formation fluids are compressible or partially so, the momentum is more difficult to determine. Consider the following development of an expression for the momentum of a compressible fluid.

From the conservation of mass, Equation 5.16:

$$\rho v A = \rho_i v_i A_i$$

And, for a gas, the mass rate of flow from Equation 5.15 is

$$\omega = \rho v A = \rho q$$

Substituting into the momentum equation gives the momentum of the gas as:

$$M = \frac{\rho q v_i}{g_c}$$

Rearranging the equation for the conservation of mass gives an expression for the velocity of the gas, v_i, at any location as follows:

$$v_i = \frac{q_i}{A}$$

$$v_i = \frac{\rho q}{\rho_i A}$$

From the Ideal Gas Law an expression for ρ_i, the density of the gas at any point in the flow stream, is given by Equation 5.18:

$$\rho_i = \frac{S_g M_a P_i}{z_i T_i R} \qquad (5.18)$$

Where:

S_g = Specific gravity of the gas

M_a = Molecular weight of air

P_i = Pressure at point i, $\dfrac{lb_f}{ft^2}$

z_i = Compressibility factor at point i

T_i = Temperature at point i, °Rankine

R = Units conversion constant

Substituting results in an expression for v_i, the velocity of the gas at point i as follows:

$$v_i = \frac{\rho q z_i T_i R}{S_g M_a P_i A} \qquad (5.19)$$

Making the final substitution gives the final expression for the momentum of the gas:

$$M = \frac{(\rho q)^2 z_i T_i R}{S_g M_a P_i g_c A} \qquad (5.20)$$

In this development, all units are BASIC! That means that these equations can be used in any system as long as the variables are entered in their basic units. In the English system, the units would be pounds, feet and seconds. In the metric system, the units would be grams, centimeters and seconds. Of course, the units conversion constants would have to be changed accordingly.

Consider the example at the Pioneer Production Company Martin No 1-7:

Example 5.6
Given:

Figure 5.12

Bottomhole pressure,	P_b	=	5000 psi
Volume rate of flow,	q	=	10 mmscfpd
Specific gravity,	S_g	=	0.60
Flowing surface pressure,	P_a	=	14.65 psia
Kill-mud density,	ρ_1	=	15 ppg
Tubing OD		=	2 3/8 inches
Tubing ID		=	1.995 inches
Casing ID		=	4.892 inches
Temperature at 4,000 feet,	T_i	=	580 °Rankine
Flowing pressure at 4,000 feet,	P_i	=	317.6 psia
Compressibility factor:			
at 4,000 feet,	z_i	=	1.00

Required:

Determine the momentum of the gas at 4,000 feet and the rate at which the kill mud will have to be pumped in order for the momentum of the kill mud to exceed the momentum of the gas and kill the well.

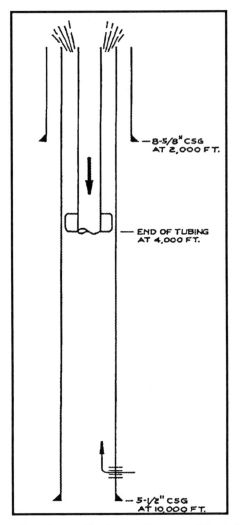

Figure 5.12 - *Pioneer Production Company - Martin No. 1-7*

Solution:

The momentum of the gas at 4,000 feet can be calculated from Equation 5.20 as follows:

$$M = \frac{(\rho q)^2 z_i T_i R}{S_g M_a P_i g_c A}$$

Substituting and in the proper units,

$$\rho \quad = 0.0458 \; \frac{lb_m}{ft^3}$$

$$q \quad = 115.74 \; \frac{ft^3}{sec}$$

$$z_i \quad = 1.00$$

$$T_i \quad = 580 \; °\text{Rankine}$$

$$R \quad = 1544 \; \frac{ft - lb_f}{°R - lb_m}$$

$$S_g \quad = 0.60$$

$$M_a \quad = 28.97$$

$$P_i \quad = 45734 \; \frac{lb_f}{ft^2}$$

$$g_c \quad = 32.2 \; \frac{lb_m - ft}{lb_f - sec^2}$$

$$A \quad = 0.1305 \; ft^2$$

$$M = \frac{[(0.0458)(115.74)]^2 (1.00)(580)(1544)}{(0.60)(28.97)(45734)(32.2)(0.1305)}$$

$$M = 7.53 \; lb_f$$

The rate at which the kill mud must be pumped can be determined by re-arranging Equation 5.17, substituting in the proper units and solving for the volume rate of flow as follows:

$$q = \left[\frac{Mg_c A}{\rho} \right]^{\frac{1}{2}}$$

Where:

$$M \quad = 7.53 \; lb_f$$

$$\rho \quad = 112.36 \frac{lb_m}{ft^3}$$

$$A \quad = 0.1305 \ ft^2$$

$$q = \left[\frac{(7.53)(32.2)(0.1305)}{112.36} \right]^{\frac{1}{2}}$$

$$q = 0.5307 \frac{ft^3}{sec}$$

$$q = 5.7 \ bbl/min$$

The foregoing example is only one simple example of the use of the Momentum Kill technology. The momentum of the wellbore fluids is more difficult to calculate when in multiphase flow. However, the momentum of each component of the flow stream is calculated and the total momentum is the sum of the momentum of each component.

References

1.   Grace, Robert D., "Fluid Dynamics Kill Wyoming Icicle." World Oil, April 1987, page 45.

2.   Blount, E. M. and Soeiinah, E., "Dynamic Kill: Controlling Wild Wells a New Way," World Oil, October 1981, page 109.

3.   Kouba, G. E., et al. "Advancements in Dynamic Kill Calculations for Blowout Wells," SPE Drilling and Completion, September 1993, page 189.

4.   Courtesy of P. D. Storts and Williford Energy.

5.   Orkiszewski, J., "Predicting Two-Phase Pressure Drops in Vertical Pipe," Journal of Petroleum Technology, June 1967, page 829.

CHAPTER SIX
SPECIAL SERVICES IN WELL CONTROL

SNUBBING

Snubbing is the process of running or pulling tubing, drillpipe or other tubulars in the presence of sufficient surface pressure present to cause the tubular to be forced out of the hole. That is, in snubbing the force due to formation pressure's acting to eject the tubular exceeds the buoyed weight of the tubular. As illustrated in Figure 6.1, the well force, F_w, is greater than the weight of the pipe. The well force, F_w, is a combination of the pressure force, buoyant force and friction force.

Stripping is similar to snubbing in that the tubular is being run into or pulled out of the hole under pressure; however, in stripping operations the force resulting from the surface pressure is insufficient to overcome the weight of the string and force the tubular out of the hole (Figure 6.2).

Snubbing or stripping operations through rams can be performed at any pressure. Snubbing or stripping operations through a good quality annular preventer are generally limited to pressures less than 2000 psi. Operations conducted through a stripper rubber or rotating head should be limited to pressures less than 250 psi. Although slower, ram-to-ram is the safest procedure for conducting operations under pressure.

Some of the more common snubbing applications are as follows:

- Tripping tubulars under pressure
- Pressure control/well killing operations
- Fishing, milling or drilling under pressure
- Completion operations under pressure

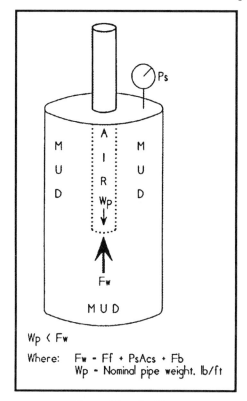

Figure 6.1 - Snubbing

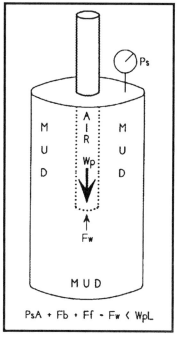

Figure 6.2 - Stripping

There are some significant advantages to snubbing operations. Snubbing may be the only option in critical well control operations. In general, high pressure operations are conducted more safely. For completion operations, the procedures can be performed without kill fluids, thereby eliminating the potential for formation damage.

There are, however, some disadvantages and risks associated with snubbing. Usually, the procedures and operations are more complex. Snubbing is slower than stripping or conventional tripping. Finally, during snubbing operations there is always pressure and usually gas at the surface.

EQUIPMENT AND PROCEDURES

The Snubbing Stack

There are many acceptable snubbing stack arrangements. The basic snubbing stack is illustrated in Figure 6.3. As illustrated, the lowermost rams are blind safety rams. Above the blind safety rams are the pipe safety rams. Above the pipe safety rams is the bottom snubbing ram, followed by a spacer spool and the upper snubbing ram. Since a ram preventer should not be operated with a pressure differential across the ram, an equalizing loop is required to equalize the pressure across the snubbing rams during the snubbing operation. The pipe safety rams are used only when the snubbing rams become worn and require changing.

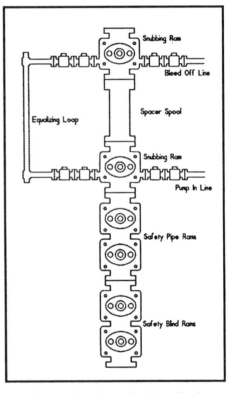

When a snubbing ram begins to leak, the upper safety ram is closed and the pressure above the upper safety ram is released through the bleed-off line. The snubbing ram is then repaired. The pump in line can be used to equalize the pressure across the safety ram and the snubbing operation continued. Since all rams hold pressure from below, an inverted ram must be included below the stack if the snubbing stack is to be tested to pressures greater than well pressure.

The Snubbing Procedure

The snubbing procedure is illustrated beginning with Figure 6.4. As illustrated in Figure 6.4 when snubbing into the hole, the tool joint or

Figure 6.3 - Basic Snubbing Stack

connection is above the uppermost snubbing ram which is closed. Therefore, the well pressure is confined below the upper snubbing ram.

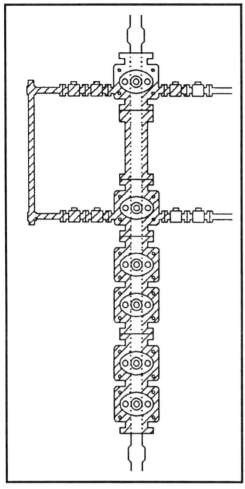

Figure 6.4 *- Snubbing into the hole*

When the tool joint reaches the upper snubbing ram, the lower snubbing ram and equalizing loop are closed, which confines the well pressure below the lower snubbing ram. The pressure above the lower snubbing ram is released through the bleed-off line as shown in Figure 6.5. After the pressure is released above the lower snubbing ram, the upper snubbing ram is opened, the bleed-off line is closed and the connection is lowered to a position immediately above the closed lower snubbing ram as illustrated in Figure 6.6. The upper snubbing ram is then closed and the equalizing loop is opened, which equalizes the pressure across the lower snubbing ram (Figure 6.7). The lower snubbing ram is then opened and the pipe is lowered through the closed upper snubbing ram until the next connection is immediately above the upper snubbing ram. With the next connection above the upper snubbing ram, the procedure is repeated.

Snubbing Equipment

If a rig is on the hole, it can be used to snub the pipe into the hole. The rig-assisted snubbing equipment is illustrated in Figure 6.8. With the stationary slips released and the traveling slips engaged, the traveling block is raised and the pipe is forced into the hole. At the bottom of the stroke, the stationary slips are engaged and the traveling slips are released. The counterbalance weights raise the traveling slips as the traveling block is lowered. At the top of the stroke, the traveling slips are engaged, the stationary slips are released and the procedure is repeated. The conventional snubbing system moves the pipe. If drilling operations under pressure are required, a power swivel must be included.

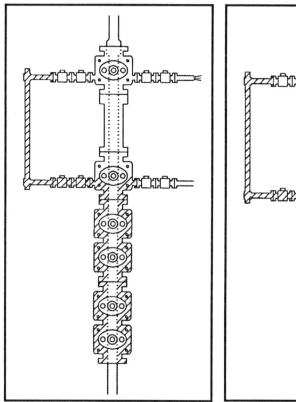
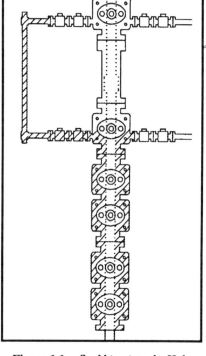

Figure 6.5 - Snubbing into the Hole   *Figure 6.6 - Snubbing into the Hole*

In the absence of a rig, a hydraulic snubbing unit can be used. A hydraulic snubbing unit is illustrated in Figure 6.9. With a hydraulic snubbing unit, all work is done from the work basket with the hydraulic system replacing the rig. The hydraulic system has the capability to circulate and rotate for cleaning out or drilling.

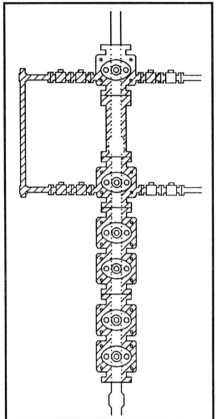

Figure 6.7 - Snubbing into the Hole

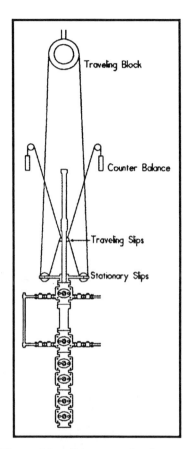

Figure 6.8 - Conventional or Rig Assisted Snubbing Unit

Theoretical considerations

As shown in Figure 6.1, snubbing is required when the well force, F_w, exceeds the total weight of the tubular. The snubbing force is equal to the net upward force as illustrated in Equation 6.1 and Figure 6.1:

$$F_{sn} = W_p L - \left(F_f + F_B + F_{wp} \right) \tag{6.1}$$

Where:

$$
\begin{aligned}
W_p &= \text{Nominal weight of the pipe, #/ft} \\
L &= \text{Length of pipe, feet} \\
F_f &= \text{Friction force, } lb_f \\
F_B &= \text{Buoyant force, } lb_f \\
F_{wp} &= \text{Well pressure force, } lb_f
\end{aligned}
$$

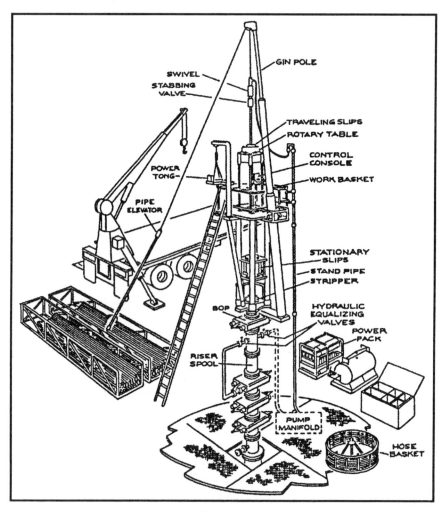

Figure 6.9

The well pressure force, F_{wp}, is given by Equation 6.2:

$$F_{wp} = 0.7854 D_p^2 P_s \qquad\qquad (6.2)$$

Where:

P_s = Surface pressure, psi

D_p = Outside diameter of tubular exposed to P_s, inches

As shown in Equation 6.2, the diameter of the pipe within the seal element must be considered. When running pipe through an annular or stripper, the outside diameter of the connection is the determining variable. When stripping or snubbing pipe from ram to ram, only the pipe body is contained within the seal elements; therefore, the outside diameter of the tube will determine the force required to push the pipe into the well. With drillpipe, there is a significant difference between the diameter of the pipe body and the tool joint. Example 6.1 illustrates the calculation of the wellhead pressure force:

Example 6.1

Given:

Surface pressure,	P_s =	1500 psi
Work string	=	4.5-inch drillpipe
Pipe OD,	D_p =	4.5 inches
Connection OD,	D_{pc} =	6.5 inches

Required:

The well pressure force when the annular is closed on

1. The tube (Figure 6.10)

2. The connection (Figure 6.11)

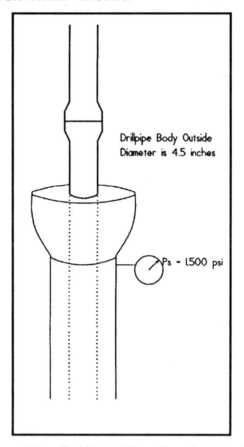

Figure 6.10 - Snubbing Drillpipe Through the Annular

Solution:

1.   When the annular is closed on the tube, the force associated with the pressure can be determined using Equation 6.2:

$$F_{wp} = 0.7854 D_p^2 P_s$$

$$F_{wp} = 0.7854(4.5^2)(1500)$$

$$F_{wp} = \mathbf{23{,}857} \ lb_f$$

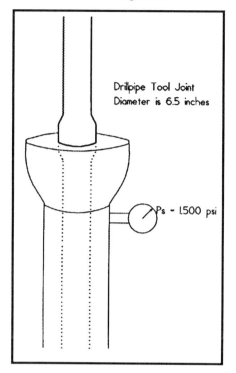

Figure 6.11 - Snubbing the Tool Joint Through the Annular

2.   When the annular is closed on a tool joint, the force is calculated using the diameter of the connection:

$$F_{wp} = 0.7854(6.5^2)(1500)$$

$$F_{wp} = 49,775 \; lb_f$$

In addition to the pressure area force, the friction force must be considered. Friction is that force which is tangent to the surface of contact between two bodies and resisting movement. Static friction is the force that resists the initiation of movement. Kinetic friction is the force resisting movement when one body is in motion relative to the other. The force required to overcome static friction is always greater than that required to maintain movement (kinetic friction). Since friction is a resistance to motion, it acts in the direction opposite the pipe movement. Friction acts upward when snubbing or stripping into a well and downward when snubbing or stripping out of a well. The magnitude of the force required to overcome friction is a function of the roughness of the surface areas in contact, total surface area, the lubricant being used and the closing force applied to the BOP.

Additional friction or drag may result between the snubbing string and the wall of the hole. In general, the larger the dogleg severity, inclination and tension (or compression) in the snubbing string, the greater the friction due to drag.

In addition to the forces associated with pressure and friction, the buoyant force affects the snubbing operation. Buoyancy is the force exerted by a fluid (either gas or liquid) on a body wholly or partly immersed and is equal to the weight of the fluid displaced by the body.

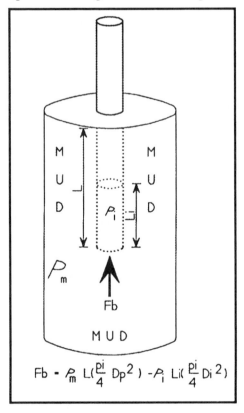

$$Fb = \rho_m L(\frac{pi}{4} Dp^2) - \rho_i Li(\frac{pi}{4} Di^2)$$

Figure 6.12 - The Buoyant Force

As illustrated in Figure 6.12, the buoyant force, F_B, is given by Equation 6.3:

$$F_B = 0.7854\left(\rho_m D_p^2 L - \rho_i D_i^2 L_i\right)$$ (6.3)

Where:

$$
\begin{aligned}
\rho_m &= \text{Mud gradient in annulus, psi/ft} \\
\rho_i &= \text{Fluid gradient inside pipe, psi/ft} \\
D_p &= \text{Outside diameter of pipe, inches} \\
D_i &= \text{Inside diameter of pipe, inches} \\
L &= \text{Length of pipe below BOP, feet} \\
L_i &= \text{Length of column inside pipe, feet}
\end{aligned}
$$

If the pipe is being snubbed into the hole dry, the density of the air is negligible and the $\rho_i D_i^2 L_i$ term is negligible. If the inside of the pipe is full or partially full, the $\rho_i D_i^2 L_i$ term cannot be ignored. If the annulus is partially filled with gas, the $\rho_m D_p^2 L$ term must be broken into its component parts. If the annulus contains muds of different densities, each must be considered. The determination of the buoyant force is illustrated in Example 6.2, and Equation 6.3 becomes

$$F_B = 0.7854\left[\left(\rho_{m1} L_1 + \rho_{m2} L_2 + \rho_{m3} L_3 + \ldots + \rho_{mx} L_x\right)D_p^2 - \rho_i L_i D_i^2\right]$$

Where:

$$
\begin{aligned}
L_1 &= \text{Column length of fluid having a density gradient } \rho_{m1} \\
L_2 &= \text{Column length of fluid having a density gradient } \rho_{m2} \\
L_3 &= \text{Column length of fluid having a density gradient } \rho_{m3} \\
L_x &= \text{Column length of fluid having a density gradient } \rho_{mx}
\end{aligned}
$$

Example 6.2

Given:

Schematic   =   Figure 6.12

Mud gradient, ρ_m =   0.624 psi/ft

Length of pipe, L =   2,000 feet

Tubular   =   4 ½-inch 16.6 #/ft drillpipe

Tubular is dry.

Required:
The buoyant force

Solution:
The buoyant force is given by Equation 6.3:

304

$$F_B = 0.7854\left(\rho_m D_p^{\,2} L - \rho_i D_i^{\,2} L_i\right)$$

With dry pipe, Equation 6.3 reduces to

$$F_B = 0.7854 \rho_m D_p^{\,2} L$$

$$F_B = 0.7854(0.624)(4.5^2)(2000)$$

$$F_B = 19{,}849 \; lb_f$$

In this example, the buoyant force is calculated to be 19,849 lb_f. The buoyant force acts across the exposed cross-sectional area which is the end of the drillpipe and reduces the effective weight of the pipe. Without the well pressure force, F_{wp}, and the friction force, F_f, the effective weight of the 2,000 feet of drillpipe would be given by Equation 6.4:

$$W_{eff} = W_p L - F_B \tag{6.4}$$

Example 6.3
   Given:
      Example 6.2

   Required:
      Determine the effective weight of the 4 ½-inch drillpipe.

   Solution:
      The effective weight, W_{eff}, is given by Equation 6.4:

$$W_{eff} = W_p L - F_B$$

$$W_{eff} = 16.6(2000) - 19849$$

$$W_{eff} = \textbf{13,351 lb.}$$

As illustrated in this example, the weight of drillpipe is reduced from 33,200 pounds to 13,351 pounds by the buoyant force.

The maximum snubbing or stripping force required occurs when the string is first started, provided the pressure remains constant. At this point, the weight of the string and the buoyant force are minimal and may generally be ignored. Therefore, the maximum snubbing force, F_{snmx}, can be calculated from Equation 6.5:

$$F_{snmx} = F_{wp} + F_f \tag{6.5}$$

   Where:
      F_{snmx} =   Maximum snubbing force, lb_f
      F_{wp} =   Well pressure force, lb_f
      F_f =   Frictional pressure force, lb_f

As additional pipe is run in the hole, the downward force attributable to the buoyed weight of the string increases until it is equal to the well pressure force, F_{wp}. This is generally referred to as the balance point and is the point at which the snubbing string will no longer be forced out of the hole by well pressure. That is, as illustrated in Figure 6.13, at the balance point the well force, F_w, is exactly equal to the weight of the tubular being snubbed into the hole. The length of empty pipe at the balance point is given by Equation 6.6:

$$L_{bp} = \frac{F_{snmx}}{W_p - 0.0408\, \rho D_p} \tag{6.6}$$

Where:

L_{bp} = Length at balance point, feet
F_{snmx} = Maximum snubbing force, lb_f
W_p = Nominal pipe weight, $\#/ft$
ρ = Mud density, ppg
D_p = Outside diameter of tubular, inches

After the pipe is filled, the net downward force is a positive snubbing force as given by Equation 6.1.

In a normal snubbing situation, the work string is run to a point just above the balance point without filling the work string. While snubbing, the well force must be sufficiently greater than the weight of the pipe to cause the slips to grip the pipe firmly. It is intended that, after the pipe is filled, the weight of the pipe is sufficient to cause the slips to grip the pipe firmly. This practice increases the string weight and reduces the risk of dropping the work string near the balance point.

The determination of the balance point is illustrated in Example 6.4:

Example 6.4

Given:
4 ½-inch 16.6 #/ft drillpipe is to be snubbed ram to ram into a well containing 12-ppg mud with a shut-in wellhead pressure of 2500 psi. The friction contributable to the BOP ram is 3,000 lb_f. The internal diameter of the drillpipe is 3.826 inches.

Required:
1.   The maximum snubbing force required

2.   Length of empty pipe to reach the balance point

3.   The net downward force after the pipe is filled at the balance point

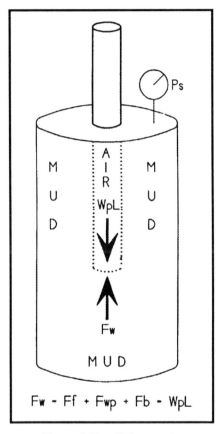

Figure 6.13 - Balance point

Solution:

1.   The maximum snubbing force is given by Equation 6.5:

$$F_{snmx} = F_{wp} + F_f$$

Combining Equations 6.5 and 6.2:

$$F_{snmx} = 0.7854 D_p^{\ 2} P_s + F_f$$

$$F_{snmx} = 0.7854(4.5^2)(2500) + 3000$$

$$F_{snmx} = \textbf{42,761 } lb_f$$

2.   The length of empty pipe at the balance point is given by Equation 6.6:

$$L_{bp} = \frac{F_{snmx}}{W_p - 0.0408 \, \rho D_p^{\ 2}}$$

$$L_{bp} = \frac{42,761}{16.60 - 0.0408(12)(4.5^2)}$$

$$L_{bp} = \textbf{6,396 feet}$$

3.   The net force after the pipe is filled is given by Equation 6.1:

$$F_{sn} = W_p L - \left(F_f + F_B + F_{wp} \right)$$

Since $F_{snmx} = F_f + F_{wp}$,

$$F_{sn} + F_{wp} = \textbf{85,958 } lb_f$$

The buoyant force, F_B, is given by Equation 6.3:

$$F_B = 0.7854 \left(\rho_m D_p^{\ 2} L - \rho_m D_i^{\ 2} L_i \right)$$

$$F_B = 0.7854 \left[(0.624)(4.5^2)(6396) - (0.624)(3.826^2)(6396) \right]$$

$$F_B = 17{,}591 \text{ lbs}$$

Therefore,

$$F_{sn} = 6{,}396(16.6) - 42{,}761 - 17{,}591$$

$$F_{sn} = 45{,}822 \text{ lbs}$$

EQUIPMENT SPECIFICATIONS

In hydraulic snubbing operations, the hoisting power required is produced by pressure applied to a multi-cylinder hydraulic jack. The jack cylinder is represented in Figure 6.14. Pressure is applied to different sides of the jack cylinder depending on whether snubbing or stripping is being done. During snubbing, the jack cylinders are pressurized on the piston rod side and on the opposite side for lifting or stripping.

Once the effective area of the jack is known, the force required to lift or snub a work string can be calculated using Equations 6.7 and 6.8:

$$F_{snub} = 0.7854 P_{hyd} N_c \left(D_{pst}^{\ 2} - D_r^{\ 2} \right) \tag{6.7}$$

$$F_{lift} = 0.7854 P_{hyd} N_c D_{pst}^{\ 2} \tag{6.8}$$

Where:

F_{snub} = Snubbing force, lb_f

F_{lift} = Lifting force, lb_f

D_{pst} = Piston diameter, inches

D_r = Outside diameter of piston rod in jack cylinder, inches

N_c = Number of active jack cylinders

P_{hyd} = Hydraulic pressure needed on jacks to snub/lift, psi

The determination of the snubbing and lifting force is illustrated in Example 6.5:

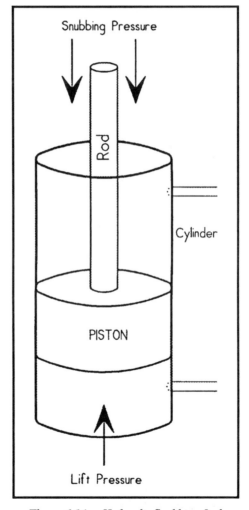

Figure 6.14 - *Hydraulic Snubbing Jack*

Example 6.5

Given:

A hydraulic snubbing unit Model 225 with four jack cylinders. Each cylinder has a 5-inch diameter bore and a 3.5-inch diameter piston rod. The maximum hydraulic pressure is 2500 psi.

Required:

1.   The snubbing force, F_{snub}, at the maximum pressure

2.   The lifting force, F_{lift}, at the maximum pressure

Solution:

1.   The snubbing force at 2500 psi is given by Equation 6.7:

$$F_{snub} = 0.7854 P_{hyd} N_c \left(D_{pst}^{\,2} - D_r^{\,2}\right)$$

$$F_{snub} = 0.7854(2500)(4)\left(5^2 - 3.5^2\right)$$

$$F_{snub} = \mathbf{100{,}139} \; lb_f$$

2.   Calculate the lifting force at 2500 psi using Equation 6.8:

$$F_{lift} = 0.7854 P_{hyd} N_c D_{pst}^{\,2}$$

$$F_{lift} = 0.7854(2500)(4)\left(5^2\right)$$

$$F_{lift} = \mathbf{196{,}350} \; lb_f$$

The hydraulic pressure required to snub or lift in the hole can be calculated by rearranging Equation 6.8.

Example 6.6 illustrates the determination of the hydraulic pressure required for a specific lifting or snubbing force:

Example 6.6

Given:

The same hydraulic snubbing unit as given in Example 6.5. The hydraulic jacks have an effective snubbing area of 40.06 in^2 and a effective lifting area of 78.54 in^2.

Required:

1. The hydraulic jack pressure required to produce a snubbing force of 50,000 lbs.

2. The hydraulic jack pressure required to produce a lifting force of 50,000 lbs.

Solution:

1. The hydraulic pressure required for snubbing is determined by rearranging Equation 6.7:

$$P_{shyd} = \frac{F_{snub}}{0.7854\left(D_{pst}^{\ 2} - D_r^{\ 2}\right)N_c} \qquad (6.9)$$

$$P_{shyd} = \frac{50000}{0.7854\left(5^2 - 3.5^2\right)4}$$

$$P_{shyd} = \frac{50000}{40.06}$$

$$P_{shyd} = 1248 \text{ psi}$$

2. The hydraulic pressure required for lifting is determined by rearranging Equation 6.8:

$$P_{lhyd} = \frac{F_{lift}}{0.7854 D_{pst}^{2} N_c} \qquad (6.10)$$

$$P_{lhyd} = \frac{50000}{0.7854(5^2)4}$$

$$P_{lhyd} = \frac{50000}{78.54}$$

$$P_{lhyd} = 637 \text{ psi}$$

Table 6.1 is a listing of the dimensions and capacity of snubbing units normally utilized.

Pipe Buckling Calculations

After determining the required snubbing force, this force must be compared with the compressive load that the work string can support without buckling. Pipe buckling occurs when the compressive force placed on the work string exceeds the resistance of the pipe to buckling. The smallest force at which a buckled shape is possible is the critical force. Buckling occurs first in the maximum unsupported length of the work string, which is usually in the window area of the snubbing unit if a window guide is not installed.

When the work string is subjected to a compressive load, two types of buckling may occur. Elastic or long-column buckling occurs along the major axis of the work string. The pipe bows out from the center line of the wellbore as shown in Figure 6.15a. Inelastic or local-intermediate buckling occurs along the longitudinal axis of the work string as shown in Figure 6.15b.

Table 6.1
Dimensions and Capacities of Snubbing Units

Model	150	225	340	600
Number of Cylinders	4	4	4	4
Cylinder Diameter (in)	4.0	5.0	6.0	8.0
Piston Rod Diameter (in)	3.0	3.5	4.0	6.0
Effective Lift Area (in^2)	50.27	78.54	113.10	201.06
Lifting Capacity at 3000 psi (lbs)	150,796	235,619	339,292	603,186
Effective Snub Area (in^2)	21.99	40.06	62.83	87.96
Snubbing Capacity at 3000 psi (lbs)	65,973	120,166	188,496	263,894
Effective Regenerated Lift Area (in^2)	28.27	38.48	50.27	113.10
Regenerated Lift Capacity at 3000 psi (lbs)	84,810	115,440	150,810	339,300
Block Speed Down (fpm)	361	280	178	137
Block Speed Up (fpm)	281	291	223	112
Bore through Unit (in)	8	11	11	14
Stroke (in)	116	116	116	168
Rotary Torque (ft-lbs)	1,000	2,800	2,800	4,000
Jack Weight (lbs)	5,800	8,500	9,600	34,000
Power Unit Weight (lbs)	7,875	8,750	8,750	11,000

As illustrated in Figure 6.16, the buckling load is a function of the slenderness ratio. In order to determine the type of buckling which may occur in the work string , the column slenderness ratio, S_{rc}, is compared to the effective slenderness ratio, S_{re}, of the work string. If the effective slenderness ratio, S_{re}, is greater than the column slenderness ratio, S_{rc} $(S_{re} > S_{rc})$, elastic or long-column buckling will occur. If the column slenderness ratio, S_{rc}, is greater than the effective slenderness ratio, S_{rc} $(S_{rc} > S_{re})$, inelastic or local-intermediate buckling will occur. The column slenderness ratio, S_{rc}, divides elastic and inelastic buckling.

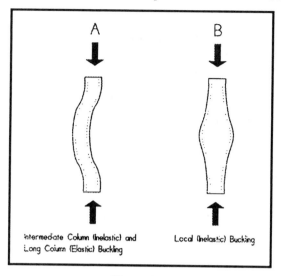

Figure 6.15

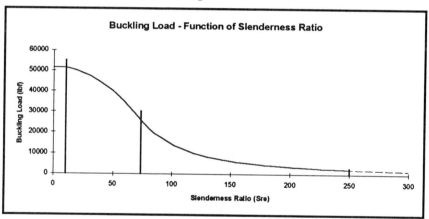

Figure 6.16

The column slenderness ratio, S_{rc}, is given by Equation 6.11:

$$S_{rc} = 4.44 \left(\frac{E}{F_y} \right)^{\frac{1}{2}}$$

(6.11)

Where:

E =   Modulus of elasticity, psi

$$F_y \quad = \quad \text{Yield strength, psi}$$

The effective slenderness ratio, S_{re}, is given by the larger result of Equations 6.12 and 6.13:

$$S_{re} = \frac{4U_L}{\left(D_p^{\,2} + D_i^{\,2}\right)^{\frac{1}{2}}} \tag{6.12}$$

$$S_{re} = \left(4.8 + \frac{D_i + t}{450t}\right)\left(\frac{D_i + t}{2t}\right)^{\frac{1}{2}} \tag{6.13}$$

Where:

$$
\begin{aligned}
U_L \quad &= \quad \text{Unsupported length, inches} \\
t \quad &= \quad \text{Wall thickness, inches} \\
D_p \quad &= \quad \text{Outside diameter of the tubular, inches} \\
D_i \quad &= \quad \text{Inside diameter of the tubular, inches}
\end{aligned}
$$

Inelastic column buckling can occur if the effective slenderness ratio, S_{re}, is less than the column slenderness ratio, S_{rc}, and is equal to or less than 250 ($S_{re} < S_{rc}$). Inelastic column buckling can occur as either local or intermediate. Whether inelastic buckling is local or intermediate is determined by a comparison of the effective slenderness ratios determined from Equations 6.12 and 6.13. If Equation 6.12 results in an effective slenderness ratio less than that obtained from Equation 6.13, local buckling occurs. If Equation 6.13 results in an effective slenderness ratio less than Equation 6.12 (but less than S_{rc}) $(S_{rc} > S_{re12} > S_{re13})$, intermediate-column buckling occurs. In either situation, a compressive load, which will cause a work string to buckle, is known as the buckling load, P_{bkl}, and is defined by Equation 6.14:

$$P_{bkl} = F_y\left(D_p^{\,2} - D_i^{\,2}\right)\left[\frac{0.7854S_{rc}^{\,2} - 0.3927S_{re}^{\,2}}{S_{rc}^{\,2}}\right] \tag{6.14}$$

for: $S_{re} < S_{rc}$ - Inelastic buckling
        $S_{re12} < S_{re13}$ - Local buckling

$S_{rc} > S_{rel2} > S_{rel3}$ - Intermediate buckling

Where:

F_y	=	Yield strength, psi
D_i	=	Inside diameter of the tubular, inches
D_p	=	Outside diameter of the tubular, inches
S_{re}	=	Effective slenderness ratio, dimensionless
S_{rc}	=	Column slenderness ratio, dimensionless

In inelastic buckling, the buckling load, P_{bkl}, can be increased by increasing the yield strength, size and weight of the work string or decreasing the unsupported section length.

Elastic (long-column) buckling is critical if the effective slenderness ratio, S_{re}, is greater than the column slenderness ratio, S_{rc}, and the effective slenderness ratio is equal to or less than 250 ($S_{re} \leq 250$). When these conditions exist, the buckling load, P_{bkl}, is defined by Equation 6.15:

$$P_{bkl} = \frac{225(10^6)(D_p{}^2 - D_i{}^2)}{S_{re}{}^2} \tag{6.15}$$

for: $S_{re} > S_{rc}$ *and* $S_{re} \leq 250$ - Long-column buckling

Under this condition, the buckling load, P_{bkl}, can be increased by decreasing the unsupported section length or increasing the size and weight of the work string. Consider the following examples:

Example 6.7
  Given:
    Work string:

	=	2 3/8 inch
	=	5.95 lb/ft
	=	P-105

Unsupported length,	U_L	=	23.5 inches
Modulus elasticity,	E	=	29×10^6 psi
Yield strength,	F_y	=	105000 psi
Outside diameter,	D_p	=	2.375 inch
Inside diameter,	D_i	=	1.867 inch
Wall thickness,	t	=	0.254 inch

Required:
The buckling load

Solution:
The column slenderness ratio is given by Equation 6.11:

$$S_{rc} = 4.44 \left(\frac{E}{F_y} \right)^{\frac{1}{2}}$$

$$S_{rc} = 4.44 \left(\frac{29(10^6)}{105,000} \right)^{\frac{1}{2}}$$

$$S_{rc} = 73.79$$

The effective slenderness ratio, S_{re}, will be the greater value as calculated from Equations 6.12 and 6.13.

Equation 6.12:

$$S_{re} = \frac{4U_L}{\left(D_p^{\,2} + D_i^{\,2}\right)^{\frac{1}{2}}}$$

$$S_{re} = \frac{4(23.5)}{\left(2.375^2 + 1.867^2\right)^{\frac{1}{2}}}$$

$$S_{re} = 31.12$$

Equation 6.13:

$$S_{re} = \left(4.8 + \frac{D_i + t}{450t}\right)\left(\frac{D_i + t}{2t}\right)^{\frac{1}{2}}$$

$$S_{re} = \left(4.8 + \frac{1.867 + 0.254}{450(0.254)}\right)\left(\frac{1.867 + 0.254}{2(0.254)}\right)^{\frac{1}{2}}$$

$$S_{re} = 9.85$$

Therefore, the correct effective slenderness ratio is the greater and is given by Equation 6.12 as 31.12.

Since S_{re} (31.12) is $< S_{rc}$ (73.79) and S_{re} is ≤ 250, failure will be in the intermediate (inelastic) mode and the buckling load is given by Equation 6.14:

$$P_{bkl} = F_y\left(D_p^{\ 2} - D_i^{\ 2}\right)\left[\frac{0.7854S_{rc}^{\ 2} - 0.3927S_{re}^{\ 2}}{S_{rc}^{\ 2}}\right]$$

$$P_{bkl} = (105000)(2.375^2 - 1.867^2)$$
$$\left[\frac{0.7854\left(73.79^2\right) - 0.3927\left(31.12^2\right)}{73.79^2}\right]$$

$$P_{bkl} = \textbf{161,907 } \textit{lb}_f$$

Consider the following example of a buckling load due to long-column mode failure:

Example 6.8
    Given:
        Work string:

	=	1 inch
	=	1.80 lb/ft
	=	P-105
Unsupported length, U_L	=	36.0 inches
Modulus of elasticity, E	=	$29 X 10^6$ psi
Yield strength, F_y	=	105000 psi
Outside diameter, D_p	=	1.315 inch
Inside diameter, D_i	=	1.049 inch
Wall thickness, t	=	0.133 inch

Required:

The buckling load

Solution:

The column slenderness ratio is calculated using Equation 6.11:

$$S_{rc} = 4.44\left(\frac{E}{F_y}\right)^{\frac{1}{2}}$$

$$S_{rc} = 4.44\left(\frac{29(10^6)}{105,000}\right)^{\frac{1}{2}}$$

$$S_{rc} = 73.79$$

The effective slenderness ratio, S_{re}, will be the greater value as calculated from Equations 6.12 and 6.13. Equation 6.12 gives

$$S_{re} = \frac{4U_L}{\left(D_p^2 + D_i^2\right)^{\frac{1}{2}}}$$

$$S_{re} = \frac{4(36)}{\left(1.315^2 + 1.049^2\right)^{\frac{1}{2}}}$$

$$S_{re} = 85.60$$

Equation 6.13 gives

$$S_{re} = \left(4.8 + \frac{D_i + t}{450t}\right)\left(\frac{D_i + t}{2t}\right)^{\frac{1}{2}}$$

$$S_{re} = \left(4.8 + \frac{1.049 + 0.133}{450(0.133)}\right)\left(\frac{1.049 + 0.133}{2(0.133)}\right)^{\frac{1}{2}}$$

$$S_{re} = 10.16$$

The greater effective slenderness ratio is given by Equation 6.12 and is 85.60.

Since S_{rc} (73.79) is $< S_{re}$ (85.60) and S_{re} is ≤ 250, failure will be in the long-column mode and Equation 6.15 will be used to determine the buckling load:

$$P_{bkl} = \frac{225(10^6)(D_p^{\ 2} - D_i^{\ 2})}{S_{re}^{\ 2}}$$

$$P_{bkl} = \frac{225(10^6)(1.315^2 - 1.049^2)}{85.60^2}$$

$$P_{bkl} = 19,309 \text{ lbs}$$

Local inelastic buckling is illustrated by Example 6.9:

Example 6.9
   Given:
      Example 6.8, except that the unsupported length, U_L, is 4 inches.

      From Example 6.8:

S_{rc} $\qquad\qquad\qquad\qquad\qquad\qquad = \quad 73.79$

S_{rel3} $\qquad\qquad\qquad\qquad\qquad\qquad = \quad 10.16$

Required:
The buckling load and mode of failure

Solution:
The slenderness ratio is given by Equation 6.12:

$$S_{rel2} = \frac{4U_L}{\left(D_p^{\,2} + D_i^{\,2}\right)^{\frac{1}{2}}}$$

$$S_{rel2} = \frac{4(4)}{\left(1.315^2 + 1.049^2\right)^{\frac{1}{2}}}$$

$$S_{rel2} = 9.51$$

Since $S_{rel2} < S_{rel3} < S_{rc}$, the buckling mode is local inelastic.

The buckling load is given by Equation 6.14:

$$P_{bkl} = F_y\left(D_p^{\,2} - D_i^{\,2}\right)\left(\frac{0.7854 S_{cr}^{\,2} - 0.3927 S_{re}^{\,2}}{S_{cr}^{\,2}}\right)$$

$$P_{bkl} = 105,000\left(1.315^2 - 1.049^2\right)$$

$$\left(\frac{0.7854\left(73.79^2\right) - 0.3927\left(10.16^2\right)}{73.79^2}\right)$$

$$P_{bkl} = 51,366 \; lb_f$$

FIRE FIGHTING AND CAPPING

Wild well fighting is more art than science. Each individual wild well fighter may have a unique approach to any problem well. However, in the past 50 years, general approaches and equipment applications have evolved. Although some unique specialty may apply to a particular situation, the general approach is to remove the remnants of the rig and well until the fire is burning through one orifice straight into the air.

Figure 6.17

EQUIPMENT

The equipment used to accomplish this task may differ. The wild well fighters from the United States rely heavily on the Athey Wagon. Two Athey Wagons are shown in Figures 6.17 and 6.18. The Athey Wagon was developed by the pipe line installation industry and adapted by the well control specialists to fire fighting operations. As illustrated, it is merely a boom on a track. The tongue of the Athey Wagon is connected to a dozer with a winch (Figure 6.19), which is used to operate and control the boom. The boom is usually about 60 feet long and the end is adapted for various operations. The hooks in Figure 6.18 are used to remove debris from around a burning well.

Figure 6.18

In order to get within working distance of the fire, all equipment is covered with corrugated, galvanized reflective metal to protect men and equipment. Heat shields and staging houses are pictured in the background of Figures 6.17 and 6.18. In addition, water is used to cool the fire and provide protection from the heat. Skid-mounted centrifugals (Figure 6.20) capable of pumping 4,800 gallons per minute pump water through conventional aluminum irrigation pipe (Figure 6.21) to sheltered fire monitors (Figure 6.22). Using the monitors for protection and staging, the monitor houses can be moved to within 50 feet of a burning well. Using the water for protection and cooling, it is then possible to work on the well with the Athey Wagon or other pieces of shielded equipment.

Figure 6.19

Figure 6.20

Figure 6.21

Figure 6.22

EXTINGUISHING THE FIRE

Some wells, such as those with toxic concentrations of hydrogen sulfide, are capped with the fire burning. In most instances, the fire is extinguished prior to the capping operation. The fire is usually extinguished with water, water in combination with fire extinguishing additives or explosives.

In many instances, several monitors are concentrated on the base of the fire and cool the fire to the extent that the fire will no longer burn. The fire-suppressing chemicals such as those found in ordinary fire extinguishers significantly increase the effectiveness of the water.

Wild well fighters are noted for the use of explosives to extinguish a fire. Generally, between 100 and 1,000 pounds, with the lower end being the most common, of dynamite are used although plastic explosives such as C_4 are also used. The dynamite is placed into a 55-gallon drum. Fire-retarding chemicals are often included. The drum is wrapped with insulating material and placed on the end of the Athey Wagon boom. The drum is positioned at the base of the plume and the dynamite is detonated. The explosion robs the fire of oxygen. The fire monitors are used to cool the area around the fire to prevent re-ignition.

CAPPING THE WELL

Once the fire is out, the capping operation begins. The well is capped on an available flange or on bare pipe, utilizing a capping stack. The capping stack is composed of one or more blind rams on top followed by a flow cross with diverted lines. The configuration of the bottom of the capping stack depends upon the configuration of the remaining well components. If a flange is available, the bottom of the capping stack below the flow cross will be an adapter flange. A flanged capping stack is illustrated in Figure 6.23. If bare pipe is exposed, the bottom of the capping stack below the flow cross will be composed of an inverted pipe ram followed by a slip ram. A capping stack with an inverted pipe ram and a slip ram is pictured in Figure 6.24. The capping stacks are placed on the well with a crane or an Athey Wagon.

Figure 6.23

Figure 6.24

In the case of exposed pipe, an alternative to the inverted pipe ram and slip ram is to install a casing flange. As illustrated in Figure 6.25, an ordinary casing flange is slipped over the exposed tubular. A crane or hydraulic jacks, supported by a wooden foundation composed of short lengths of 4 x 4's are used to set the slips on the casing head. Concrete is then poured around the jacks and foundation to the bottom of the casing head. Once the casing head is set, the excess casing is cut off using a pneumatic cutter. A capping stack can then be nippled up on the casing flange.

Figure 6.25

Once the stack is landed, the vent lines are connected and the blind ram is closed, causing the flow to be vented to a pit which should be at least 300 feet from the wellhead. With the well vented, the capping operation is complete and the control and killing operation commences.

FREEZING

Freezing is a very useful tool in well control. Invariably, the top ball valve in the drill string will be too small to permit the running of a plug. In order to remove the valve with pressure on the drillpipe, the drillpipe would have to be frozen. A wooden box is constructed around the area to be frozen. Then, a very viscous mixture of bentonite and water is pumped into the drillpipe and spotted across the area to be frozen. Next, the freeze box is filled with dry ice (solid carbon dioxide). Nitrogen should never be used to freeze because it is too cold. The steel becomes very brittle and may shatter when impacted. Several hours may be required to obtain a good plug. The rule of thumb is one hour for each 1 inch in diameter to be frozen. Finally, the pressure is bled from above the faulty valve; it is removed and replaced and the plug is permitted to thaw. Almost everything imaginable has been frozen including valves and blowout preventers.

HOT TAP

Hot tapping is another useful tool in well control. Hot tapping consists of simply flanging or saddling to the object to be tapped and drilling into the pressure. Almost anything can be hot tapped. For example, an inoperable valve can be hot tapped or a plugged joint of drillpipe can be hot tapped and the pressure safely bled to the atmosphere. In other instances, a joint of drillpipe has been hot tapped and kill fluid injected through the tap.

CHAPTER SEVEN
RELIEF WELL DESIGN AND OPERATIONS

The industry has long considered the relief well option as a last resort in well control. The problems are obvious. Even with the best surveying techniques, the bottom of the hole was unknown to any degree of certainty. The ability to communicate with the bottom of the hole was very limited and most often governed by the principle of trial and error. However, relief well technology has advanced in the past 10 years to the point that a relief well is now a viable alternative. Modern technology has made intercepting the blowout a certainty and controlling the blowout from the relief well a predictable engineering event.

HISTORY

ULSEL AND MAGNETIC INTERPRETATION INTRODUCED

On 25 March, 1970, a blowout occurred at the Shell Oil Corporation Cox No. 1 at Piney Woods, Rankin County, Mississippi.[1] The well had been drilled into the Smackover at a total depth of 21,122 feet and cased to 20,607 feet. The well flowed at rates estimated between 30 and 80 million standard cubic feet of gas per day plus 14,000 to 20,000 barrels of water per day. The hydrogen sulfide concentration in the gas stream made the gas deadly toxic to humans and, combined with the saline, produced water deadly corrosive to steels. Shortly after the well kicked on the morning of the disaster, the blowout preventer stack rose and fell over, releasing a stream of gas and invert oil-emulsion mud. Within minutes the well ignited and the derrick fell. The well had cratered.

This combination of events and circumstances made surface control at the Cox No. 1 impossible. Therefore, an all-out effort was made to control the blowout from a relief well. A conventional relief well,

Cox No. 2, was spudded on 3 May, 1970, at a surface location 3,500 feet west of the blowout. This well was designed to be drilled straight to 9,000 feet and from there directionally to 21,000 feet to be bottomed close to the Cox No. 1.

There was considerable skepticism about the effectiveness of this approach. Was directional drilling possible at these depths? Could the bottom of the blowout be determined with any reasonable accuracy? Could solids-laden fluid be communicated to the blowout through the Smackover with its relatively low porosity and permeability? Because of these uncertainties, a special task force was formed to explore new techniques. On 16 May, 1970, the Cox No. 4 was spudded 1,050 feet east of the blowout. Its mission was to intercept the blowout in the interval between 9,000 and 13,000 feet and effect a kill. The resulting work was to pioneer modern relief well technology.

Two methods of evaluation were introduced and developed and formed the basis for modern technology. One technique involved the use of resistivity measurements to determine the distance between the relief well and the blowout. The use of electrical logs for locating various drillpipe and casing fishes in wellbores was well known and many examples could be cited. However, little effort had been expended to utilize resistivity devices as a direct means of determining distance between wells. Also, there were very few examples of intersecting wellbores. Due to the nature and depth of the reservoir rocks, a wellbore intercept at or about 10,000 feet was necessary to effect a kill at the Cox No. 1.

In response to this disaster, Shell and Schlumberger developed and reported the use of ultra-long, spaced electrical logs, commercially known as ULSEL, to determine the distance between wells. The ULSEL technology was developed primarily for the profiling of salt structures or other resistive anomalies. The ULSEL tool is merely the old short-normal technique with electrode spacings of 150 feet for AM and 600 feet for the AN electrodes. An example of induction-electric log response in an intersecting well is shown in Figure 7.1. Rather inaccurate estimates of distances between wellbores are made by curve fitting techniques designed to model the short-circuiting effect of the casing in the blowout. At the Amoco R. L. Bergeron No. 1 in the Moore-Sams Field near Baton Rouge, Louisiana, which blew out in early 1980, the ULSEL technology predicted the blowout to be 12 to 18 feet from the relief well at 9,050 feet measured depth. In reality, the blowout was approximately 30 feet from the relief well.[2]

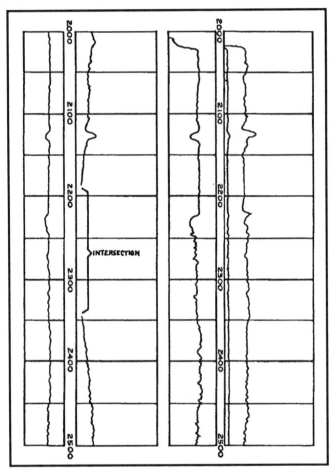

Figure 7.1 - Short Circuit Effect of Casing Ulsel Response

A most interesting technique was developed by Shell Development Company and reported by J. D. Robinson and J. P. Vogiatzis. This technique involved the utilization of sensitive magnetometers to measure the earth's magnetic field and the distortion of that magnetic field resulting from the presence of casing or other tubulars possessing remnant magnetization. Measurements were made from the relief well and sidetracks, and models were assumed in an effort to match the relative position of the two wells. These efforts were rewarded when the relief well intercepted the Cox No. 1 at approximately 10,000 feet at a position estimated to be between 3 and 9 inches from the position of the axis predicted by the calculations.[2] As a result, United States Patent #3,725,777 was awarded in recognition of this new advance in technology.

THE CONTRIBUTION OF SCHAD

A very significant and heretofore relatively unnoticed contribution was made by Charles A. Schad and is documented in United States Patent #3,731,752 filed 25 June, 1971 and issued 8 May, 1973.[3] Schad developed a highly sensitive magnetometer consisting of at least one pair of generally rectangular core elements having square hypothesis loops for use in a guidance system for off-vertical drilling. Magnetometers were well known prior to this work. However, their usefulness was limited to relatively strong magnetic fields. In wellbore detection, a sensitivity of 0.05 gammas would be required and would need to be distinguished from the horizontal component of the earth's magnetic field, which is between 14,000 and 28,000 gammas. Obviously, the development of a magnetometer with sensitivity such that the earth's magnetic field could be nulled in order to study a small magnetic field caused by a ferromagnetic source within the earth was a tremendous contribution to relief well technology. It is this magnetometer concept that is used in modern wellbore proximity logging. In addition, Schad envisioned placing a magnetic field in one wellbore and using his magnetometer to guide a second wellbore to an intercept with the first. This technology later proved to be a pioneer to relief well operations.

MAGRANGE DEVELOPED

Unfortunately, no commercial service resulted from the work of Schad or Robinson. Schlumberger continued to offer the ULSEL technology; however, the distances were subject to interpretation and no concept of direction was available. Therefore, when Houston Oil and Minerals experienced a blowout in Galveston Bay in 1975, reliable wellbore proximity logging services were not commercially available. As a result, Houston Oil and Minerals Corporation commissioned Tensor, Inc. to develop a system for making such measurements.

In response to that need and commission, Tensor developed the MAGRANGE service. The MAGRANGE service is based on United States Patent #4,072,200 which was filed 12 May, 1976 and issued to Fred J. Morris, et. al. on 7 February, 1978.[4] The Morris technology was similar to that of Robinson and Schad in that highly sensitive magnetometers were to be used to detect distortions in the earth's magnetic field caused by the presence of remnant magnetism in a ferromagnetic body.[5] However, it differed in that Morris envisioned measuring the change in magnetic gradient along a wellbore. It was reasoned that the magnetic gradient of the earth's magnetic field is small and uniform and could be differentiated from that gradient caused by a ferrous target in the blowout wellbore. The MAGRANGE service then made a continuum of measurements along the wellbore of the relief well and analyzed the change in gradient to determine the distance and direction to the blowout. This technology was state of the art for several years following the blowout at Galveston Bay and was used in many relief well operations. However, interpretation of the data proved less reliable than needed for accurate determination of the distance and direction to a blowout. In addition, detection was limited to approximately 35 feet.

WELLSPOT DEVELOPED

In early 1980, the R. L. Bergeron No. 1 was being drilled by Amoco Production Co. as a Tuscaloosa development well in the Moore-Sams Field near Baton Rouge, Louisiana, when it blew out at a depth of 18,562 feet.[9] Systematic survey errors and the limited depth of reliable investigation of the available commercial borehole proximity logs prompted the operator to seek alternate techniques. To that end, the operator contacted Dr. Arthur F. Kuckes, professor of physics at Cornell University in Ithaca, New York. In response to that challenge, new technology was developed that provided reliability never before available in relief well operations. This technology is currently marketed by Vector Magnetics, Inc. under the trade name WELLSPOT. The theoretical aspects are fully described in the referenced material.[6] The approach is quite simple and straightforward. As illustrated in Figure 7.2, an electrode is run on a conventional electric line 300 feet above a tool consisting of four magnetometers. Two AC magnetometers respond to the two components of an AC magnetic field perpendicular to the axis of the tool, and two fluxgates measure the two components of the earth's magnetic field perpendicular to the tool axis. The fluxgates act as a magnetic compass so that the tool's orientation can be determined.

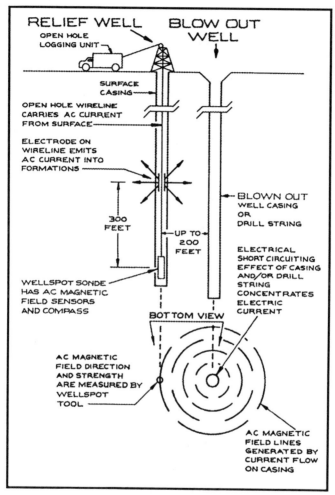

Figure 7.2 - Wellspot Operations

Between the tool and the wireline from the logging truck, there is an insulating bridle approximately 400 feet long. On this bridle cable 300 feet above the tool, an electrode emits AC electric current into the formations. In the absence of a nearby ferrous material, such as the casing in the blowout, the current flows symmetrically into the ground and dissipates. In the presence of a ferromagnetic body, such as the casing or drill string in the blowout, the flow of electric current is short-circuited, creating a magnetic field. The intensity and direction of the magnetic field are measured at the tool. The various parameters are analyzed using generally routine and theoretically straightforward mathematical analysis. Depending on the conductivity of the formations in the wellbore relative to the conductivity of the ferromagnetic body, the method can detect the blowout from the relief well at distances greater than 200 feet. As will be discussed, the method is more accurate as the relief well approaches the blowout. Typical WELLSPOT data are illustrated as Figures 7.3 and 7.4.

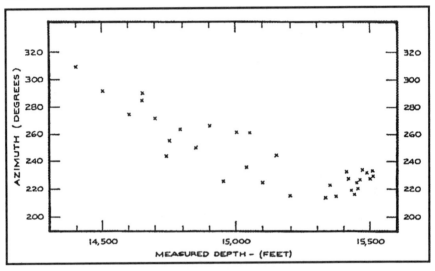

Figure 7.3 - *Wellspot Blowout Azimuth Direction from Relief Well*

MAGRANGE AND WELLSPOT COMPARED

In June 1981, Apache Corporation completed the Key 1-11 in the Upper Morrow at 16,000 feet in eastern Wheeler County of the Texas Panhandle.[7.] The Key 1-11 was one of the best wells ever drilled in the Anadarko basin, having 90 feet of porosity in excess of 20%. The original open flow potential was in excess of 90 mmscfpd. On Sunday afternoon, 4 October, 1981, after being shut in for 78 days waiting on pipeline connection, the well inexplicably erupted. The blowout that was known as the biggest in the history of the state of Texas was controlled on 8 February, 1983 when the Key 3 relief well intercepted the 5-inch liner in the blowout at 15,941 feet true vertical depth (TVD) — the deepest intercept in the history of the industry. The intercept was made possible by the alternating current - active magnetic technology.

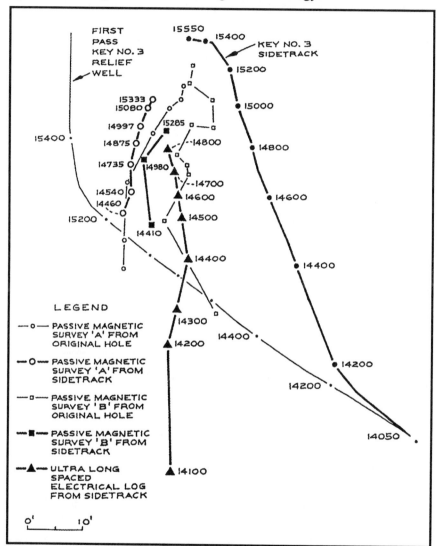

Figure 7.4 - Comparison of Various Proximity Logging Techniques

The passive magnetic technology of MAGRANGE was utilized first at the Key to direct the relief well effort. The early MAGRANGE interpretations conflicted with available data concerning the bottomhole location of the blowout. However, operations continued pursuant to the

MAGRANGE interpretations and, at 15,570 feet measured depth, the relief well was discovered to be so far off course that the relief well had to be plugged back to 14,050 feet measured depth and re-drilled. At a measured depth of 15,540 feet in the sidetrack, the MAGRANGE interpretation again conflicted with other available data. Based on the MAGRANGE interpretation, the recommendation was made to turn the relief well south to intercept the blowout (Figure 7.4). Realizing that a mistake would result in yet another sidetrack, the crew employed alternate techniques.

One alternative was WELLSPOT. The WELLSPOT interpretation was that the blowout was not 12 feet south as interpreted by MAGRANGE but 1.5 feet south. Further, it was interpreted from WELLSPOT data that the blowout would pass to the north of the relief well within the next 60 feet. Obviously, both interpretations could not be correct. If the WELLSPOT interpretation was correct and the relief well was turned to the south as recommended from the MAGRANGE interpretation, a plug back would be required. If the MAGRANGE interpretation was correct and the relief well was not turned to the south, a side track would result. In order to resolve the conflict, 60 feet of additional hole were made and both surveys were rerun. Both interpretations agreed that the blowout wellbore had passed from south to north of the relief well wellbore, confirming the WELLSPOT interpretation. Both techniques were used throughout the remainder of the relief well operation at the Apache Key. Neither was given benefit of the other's interpretation prior to offering its own interpretation. The MAGRANGE and WELLSPOT interpretations conflicted in every aspect except direction at each subsequent logging point. Without exception, the WELLSPOT interpretation was proven to be correct.

RELIABILITY OF PROXIMITY LOGGING

How reliable is wellbore proximity logging? This question was investigated at the TXO Marshall relief well operation.[8] In this case, the first indications of the blowout wellbore were received at the relief well wellbore while the blowout was 200 feet away. The relief well intercepted the blowout at 13,355 feet. The plan view from interpretation of a gyro previously run in the blowout was utilized to evaluate the accuracy of the

WELLSPOT interpretations.   Table 7.1 and Figure 7.5 summarize the proximity logging runs.   For the early runs, the blowout was about 30% further away than the interpretations predicted although the direction given was always correct.

Table 7.1 - Proximity Logging Summary

Run No.	Date	Calculated distance, ft A-5 to A-1	Calculated direction, ° A-5 to A-1	Depth, ft MD	TVD	Actual distance ft A-5 to A-1
1	Aug. 18	90 ± 15	318 ± 6	12,000	11,957	138
2	Aug. 22	50 ± 10	319 ± 4	12,112	12,068	122
3	Aug. 28	50 ± 10	317 ± 3	12,212	12,168	109
4	Aug. 31	65 ± 20	318 ± 4	12,341	12,295	94
5	Sept. 4	68 ± 10	319 ± 4	12,488	12,442	78
6	Sept. 8	48 ± 7	320 ± 5	12,625	12,578	65
7	Sept. 11	32 ± 7	319 ± 4	12,771	12,723	48
8	Sept. 15	26 ± 7	320 ± 4	12,884	12,836	37
9	Sept. 18	24 ± 5	316 ± 4	13,012	12,963	27
10	Sept. 23	11 ± 2	309 ± 4	13,142	13,093	17
11	Sept. 29	11 ± 2	285 ± 4	13,248	13,199	14
12	Oct. 2	6.5 ± 1.5	292 ± 4	13,300	13,250	8.5
13	Oct. 5	2 ± 0.75	295 ± 5	13,360	13,310	2.5

The greatest discrepancy occurred on the second run at 12,068 feet TVD. Pursuant to the plan view, the relief well was 122 feet away from the blowout as opposed to the 50 feet predicted from the proximity log interpretation. It was not until the ninth run and the wells had closed to within 30 feet that the two wellbores were actually within the prescribed error limits of the interpretation (see Figure 7.6). Further, two of the remaining four interpretations were in error in excess of the prescribed error limits. Interpretation is continuing to improve. However, the current state of the art dictates that the wellbore proximity log should be run frequently to insure the intercept, particularly as the wellbores converge. At the Marshall, for example, 13 surveys were run in 1,336 feet of drilled hole for an average of approximately 100 feet per survey. Given the present state of the art, survey intervals should not exceed 200 feet in order to accomplish a wellbore intercept. In the very near future, wellbore proximity logging will be offered, which operates like a steering tool or a measurement while drilling tool (MWD).

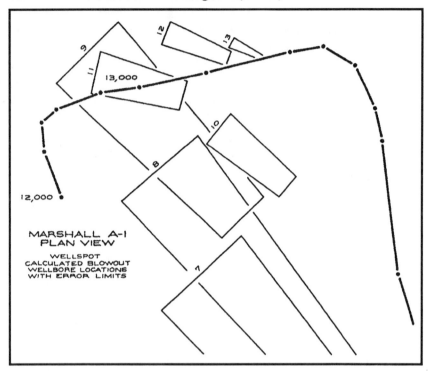

Figure 7.5

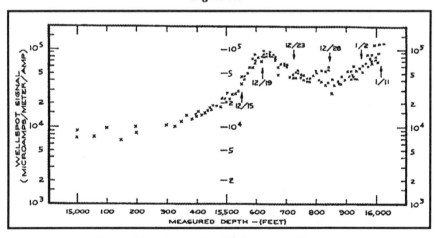

Figure 7.6 - Bottom Hole Wellspot Signal Strength

RELIABILITY OF COMMERCIAL WELLBORE SURVEY INSTRUMENTS

The accuracy and reliability of survey instruments are of great interest in relief well operations. For various reasons, sometimes the relief well has to be plugged back and re-drilled after contact has been made with the blowout. With what degree of accuracy, reliability and repeatability can the relief well be plugged back and drilled to the position of the blowout as described by available survey equipment?

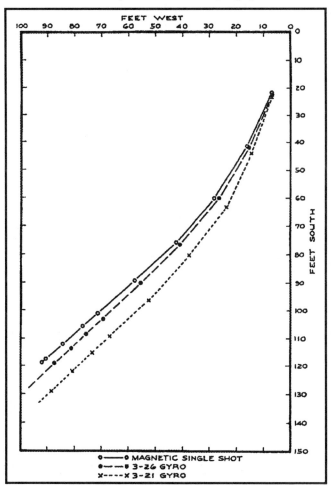

Figure 7.7 - *Marshall A-1X Plan View At 12,600 Feet Relative to 11,800 Feet*

Several directional surveys were available at the TXO Marshall. Unfortunately, there were serious discrepancies among them. As illustrated in Figure 7.7, two north-seeking gyro surveys (3-21 and 3-26) and a magnetic single-shot were tied together at 11,800 feet and plotted. The two gyro surveys, which represent the state of the art as of this writing, were very discordant with one another even though they were run with the SAME TOOL, by the SAME OPERATOR and using the SAME WIRELINE. After only 650 feet MD of surveying from 11,800 feet to 12,450 feet, the north-seeking gyros disagreed by 7 feet for an overall reliability of only 11 feet per 1,000 feet. Further, the 3-21 gyro survey disagreed with the single-shot data by 10 feet in the 650 feet for an overall reliability of 15.7 feet per 1,000 feet. The deviation in this portion of the hole varied between 10 and 14 degrees. These values for reliability and repeatability are far worse than those normally quoted within the industry.

Survey data were also available from other portions of the Marshall for comparison. In the upper part of the hole between 10,650 feet and 11,500 feet, there were four sets of survey data: a magnetic single-shot, a magnetic multi-shot (1-17), and the two north-seeking gyros (3-21 and 3-26). These data were plotted together at 10,650 feet and their deviations are illustrated as Figure 7.8. Analysis of Figure 7.8 indicates similar systematic relationships between the two gyro runs and the magnetic single-shot data. The magnetic multi-shot results seem to support the 3-21 gyro data. The maximum discrepancy is between the magnetic single-shot and the 3-26 gyro. As illustrated, these two surveys disagree by approximately 9 feet over the 850-foot interval for an overall reliability of 10.6 feet per 1,000 feet.

In some instances, magnetic data compared better than the more expensive north-seeking gyro data. For example, the 1-17 magnetic multi-shot and the magnetic single-shot data were compared in the interval between 11,400 feet and 12,300 feet (see Figure 7.9). A 5-foot uncertainty is indicated over the 863-foot interval for a reliability of 5.8 feet per 1,000 feet.

Several other comparisons were made and are presented in reference 9. Careful study of the survey data obtained in this instance makes it difficult to ascribe a survey precision of better than 10 feet of lateral movement per 1,000 feet of survey. Further, since the survey errors were primarily systematic, there was no reason to prefer the more expensive north-seeking gyroscopic survey over the more conventional magnetic survey. Typical north-seeking gyros are the Seeker, Finder and Gyro Data gyros. Typical MWD magnetic instruments are Teleco and Anadril.

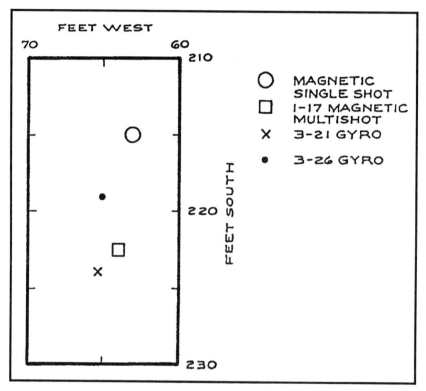

Figure 7.8 - Apparent Marshall A-1X Well Location at 11,500 feet relative to 10,650 feet

Each relief well operation is different and must be evaluated on its merits. However, in most cases, magnetic survey instruments are more appropriate for relief well operations than the more expensive north-seeking gyros. The MWD systems are more economical and expeditious during the directional operations. Further, as the well gets deeper, the MWD system offers several advantages. Surveys can record every joint, if necessary, to monitor deviation, direction and bottomhole assembly behavior. Another significant advantage is that it is not necessary to wait idly for long periods without moving the drill string while a survey is run on a slick or braided wireline.

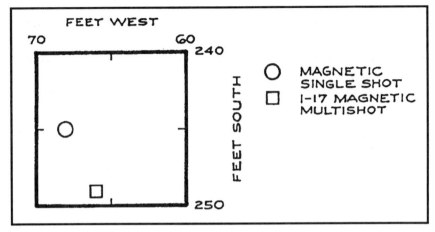

Figure 7.9 - *Apparent Marshall A-1X Well Location at 12,300 feet (Old Hole) relative to 11,400 feet*

SUBSURFACE DISTANCE BETWEEN RELIEF WELL AND BLOWOUT

A critical question in relief well operations is how close to the blowout the relief well has to be. Prior to the previously discussed technology, the position of the relief well relative to the blowout was only a poorly educated guess. Consequently, successful relief well operations could not be assured. The standard procedure was to drill into the zone that was being produced in the blowout, manifold all the pumps in the world together and pump like hell. Sometimes it worked and sometimes it didn't.

A fire broke out at Shell Oil Co.'s Platform "B" at Bay Marchand, offshore Louisiana, on December 1, 1970.[9] Of the 22 completed wells, 11 caught fire. Subsequent operations to control the blowouts were reported by Miller.[10] Production was from approximately 12,000 feet with an initial reservoir pressure of 6000 psia. The porosity in the reservoir was 29% with a permeability of 400 millidarcies. Conventional directional surveys were available on all the wells. The distances between wells were determined by analyzing the conventional directional surveys and by running ULSEL logs in the relief wells. A simplified reservoir model was used to predict performance of the relief well control operations. Eleven relief wells were drilled. The distance between the relief well and the blowout varied between 12 feet and 150 feet. In each instance the blowout zone was penetrated and water was pumped through the producing interval. Of the 11, 4 compared favorably with predicted performance. The relief wells were 12 to 18 feet from the blowouts. Water volumes to establish communication ranged from 450 barrels to 1,340 barrels. One well was killed with water while the other three required as much as 1,300 barrels of mud.

A second four compared less favorably but experienced no major problems in killing. The distances varied from 13 feet to 82 feet. Water volumes varied between 1,250 barrels and 13,000 barrels. All four wells required mud with the volumes varying between 1,187 barrels and 3,326 barrels.

The remaining three wellbores had unfavorable comparisons with predicted performance. The water volumes exceeded 100,000 barrels. In one instance, pumping into the relief well resulted in no noticeable effect on the blowout and control had to be regained from the surface.

The Bay Marchand operation was one of the more successful efforts. Often when an operation was successful, the success was due to as much luck as skill. The reasons are fairly obvious. Only the most prolific reservoir rocks demonstrate sufficient permeability to permit the flow of kill fluids containing mud solids of every imaginable size, including large quantities of barite guaranteed to be approximately 44 microns in size. If the zone fractured, there was no good reason for the fracture to extend to the blowout. The blowout had to be killed with water before kill mud could be pumped. If the kill mud was unsuccessful in controlling the blowout, communication with the blowout was usually lost because of kill-mud gel strengths and barite settling. Under these conditions, communication might not be regained.

As wells were drilled deeper, the reservoir parameters, such as permeability and porosity, were even less favorable in terms of their ability to permit the flow of solids-laden kill fluids. In wells below 15,000 feet, it is not reasonable even to think about pumping kill fluids through the formations.

With the current technology, intercepting the wellbore is the preferred approach. With an interception, killing the blowout becomes predictable. Further, the predictability permits more precise design of the relief well. Kill rates and pressures can be determined accurately and tubulars can be designed to accomplish specific objectives. With an intercept, it is no longer necessary to manifold all the pumps in captivity, as many blowouts can be controlled with the rig pumps. At the Apache Key, for example, approximately 100 barrels of mud per hour were being lost to the blowout from the relief well, and control was instantaneous. With a planned intercept, it is no longer necessary to drill the largest diameter hole imaginable in order to pump large volumes of kill fluid at high rates.

Perforating between wellbores has regained interest with the introduction of tubing-conveyed perforating guns. There are very few charges available for conventional wireline perforating of sufficient size to perforate meaningful distances. In addition, orientation can be shown mathematically to be difficult depending on the size of the casings involved and the distance between casing. However, using tubing-conveyed perforating guns, large charges can be run and oriented. Such a system was successfully used at Corpoven's Tejero blowout, northeast Venezuela. In that instance, it was reported that three attempts were required to communicate the relief well with the blowout. In the first attempt, two 6-inch guns with 14 300-gram charges in each gun were run. The shots were aligned along the gun in three rows with a displacement between each row of 5 degrees. The two guns were connected with the center rows displaced 10 degrees. On the second attempt, two 4 5/8-inch oriented TCP guns were run with seventy-two 27-gram charges. The center row was displaced 33 degrees. The relief well was believed to be within 2 feet of the blowout.

With a planned wellbore intercept, it is no longer mandatory that the relief well be drilled into the blowout zone. Other intervals in the blowout wellbore may offer more attractive targets. At both the Apache Key and the TXO Marshall, the intercepts were accomplished above the producing formations. At the Shell Cox, the intercept was accomplished approximately midway between the surface and total depth. At a recent North Sea operation, the intercept was in open hole approximately 100 feet below the end of the bit. At the Amerada Hess - Mil-Vid #3, the drill string was intercepted several hundred feet above the blowout and the relief well proceeded in direct contact with the drill string in the blowout until the final bridge was drilled and the blowout was killed. In that instance, less than 20 barrels of mud controlled the blowout. A planned intercept offers significant advantages.

SURFACE DISTANCE BETWEEN RELIEF WELL AND BLOWOUT

The distance at the surface between the relief well and the blowout is a function of overall project management; however, the closer the better. The cost of relief well operations is exponential with displacement. However, some projects are managed in such a fashion that the relief well project manager has no choice but to put the relief well a mile away in a bad direction. Ideally, overall project management will permit the relief well to be drilled within 1,000 feet of the blowout with deviation angles below 15 degrees. The best direction is that which takes advantage of the regional drift and fracture orientation tendencies.

SUMMARY

There are many other considerations in relief well operations that are related to overall project management. Large sums of money can be saved by the operator if overall management is considered. The relief well operations need to be coordinated with the total control effort. This will affect relief well location, for example, which can make or break an operation. Also, a relief well operation is not just a directional operation. It is a drilling operation, a well control operation and a logging operation as well as a directional operation. Needless to say, these considerations can impact the total cost significantly.

In summary, relief well technology has advanced to the extent that relief well operations are now a viable, reliable alternative in well control operations and should be considered in the overall planning and management of a blowout. A recent blowout at a deep, high-pressure well in the North Sea is a good example. More than a year of expensive surface work failed to provide a solution to the problem. After many expensive months, the blowout was finally controlled from the relief well. There is a good chance that the relief well would have been just as successful in the first 60 days of the operation.

RELIEF WELL PLAN OVERVIEW

1. Prepare a wellbore schematic pursuant to Figure 7.10.

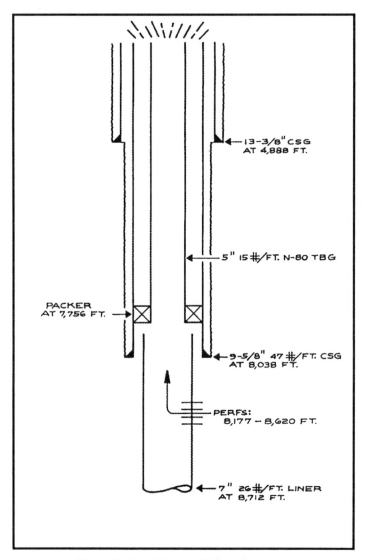

Figure 7.10 - Blowout Wellbore Schematic

2. Determine the approximate bottomhole location and preferential direction of drift as illustrated in Figure 7.11.

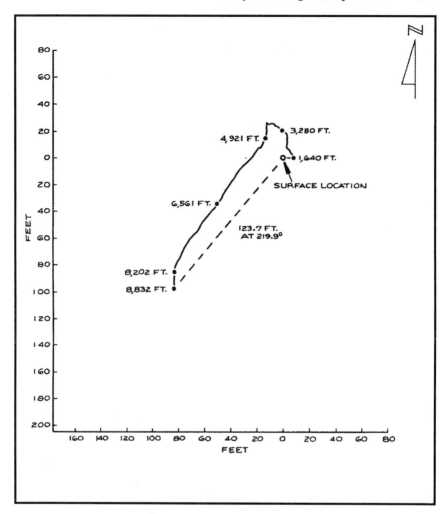

Figure 7.11 - Blowout Plan View

3. Select and build a location for the relief well approximately 820 feet northeast of the blowout.
4. Drill 17 ½-inch hole to 3,933 feet true vertical depth and reduce hole size to 12 ¼-inches for directional work.
5. Pick up motor, 1 ½ degree bent sub and MWD and build angle to 15 degrees at approximately 1.5 degrees per 100 feet.
6. Lay down motor, bent sub and MWD and pick up holding assembly.
7. Drill to intermediate casing point at 4,888 feet true vertical depth, holding angle at 15 degrees.

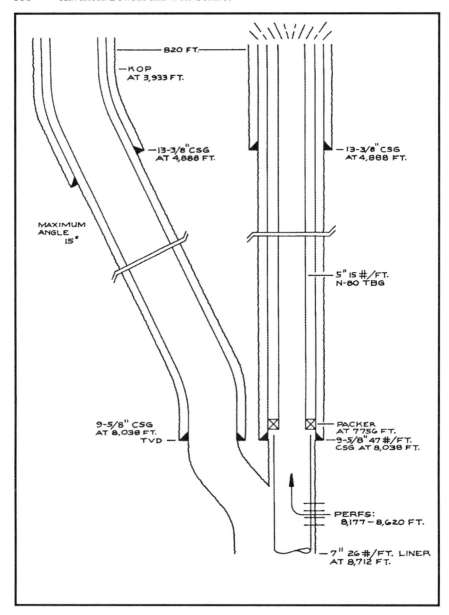

Figure 7.12

8. Open 12 ¼-inch hole to 17 ½ inches.
9. Set and cement 13 3/8-inch casing. If kill considerations permit, reduce hole size to 12 1/4-inches and casing size to 9 5/8-inches.

10. Hold 15 degrees in the 12 ¼-inch hole to 6,561 feet measured depth.
11. Run wellbore proximity log to detect blowout wellbore.
12. Make course corrections as required by the wellbore proximity log using motor, bent sub and MWD.
13. Drill to 7,090 feet measured depth and run wellbore proximity log.
14. Make course corrections as required by the wellbore proximity log using motor, bent sub and MWD.
15. Using pendulum bottomhole assembly, drop angle to vertical. Run wellbore proximity log at 200-foot intervals until blowout wellbore is confirmed.
16. Intercept the casing in the blowout well at 8,000 feet true vertical depth.
17. Drill to casing point at 8,038 feet true vertical depth and run the wellbore proximity log.
18. Run 9 5/8-inch casing to 8,038 feet true vertical depth.   If kill considerations permit, reduce hole size to 8 1/2-inches and casing size to 7 inches.
19. Drill out with 8 ½-inch (or 6-inch) bit and run wellbore proximity log as required to intercept the blowout wellbore.
20. Drill into the blowout wellbore (see Figure 7.12).
21. Kill the blowout pursuant to plan and procedure.

References

1. Robinson, J. D. and Vogiatzis, J. P., "Magnetostatic Methods for Estimating Distance and Direction from a Relief Well to a Cased Wellbore," Journal of Petroleum Technology, June 1972, pages 741 - 749.

2. Warren, Tommy M., "Directional Survey and Proximity Log Analysis of a Downhole Well Intersection," Society of Petroleum Engineers Number 10055, October 1981.

3. Schad, Charles A., United States Patent #3,731,752 issued 8 May, 1973, "Magnetic Detection and Magnetometer System Therefore."

4. Morris, F. J., et al. United States Patent #4,072,200 issued 7 February, 1978, "Surveying of Subterranean Magnetic Bodies from an Adjacent Off-Vertical Borehole."

5. Morris, F. J., et al., "A New Method of Determining Range and Direction from a Relief Well to a Blowout," SPE 6781, 1977.

6. Kuckes, Arthur F., United States Patent #4,372,398 issued 8 February, 1983, "Method of Determining the Location of a Deep-Well Casing by Magnetic Field Sensing".

7. Grace, Robert D., et al. "Case History of Texas' Largest Blowout Shows Successful Techniques on Deepest Relief Well," Oil and Gas Journal, 20 May, 1985, page 68.

8. Grace, Robert D., et al. "Operations at a Deep Relief Well: The TXO Marshall", SPE 18059, October 1988.

9. Lewis, J. B., "New Uses of Existing Technology for Controlling Blowouts: Chronology of a Blowout Offshore Louisiana," Journal of Petroleum Technology, October 1978, page 1473.

10. Miller, Robert T. and Clements, Ronald L., "Reservoir Engineering Techniques Used to Predict Blowout Control During the Bay Marchand Fire," Journal of Petroleum Technology, March 1972, page 234.

CHAPTER EIGHT
THE UNDERGROUND BLOWOUT

An underground blowout is defined as the flow of formation fluids from one zone to another. Most commonly, the underground blowout is defined by a lack of pressure response on the annulus while pumping on the drillpipe or by a general lack of pressure response while pumping. The underground blowout can be most difficult, dangerous and destructive. It can be most difficult because the conditions are hidden and can evade analysis. Often, the pressures associated with an underground blowout are nominal, resulting in a false sense of security.

It can be dangerous because some associate danger with sight. In many instances there is no physical manifestation of the underground blowout. If a well is on fire or blowing out at the surface, it commands respect. However, if the same well is blowing out underground, it is more easily ignored. Since the underground blowout is not seen, it is often not properly respected.

If the underground blowout is within 3,000 to 4,000 feet of the surface, there is the possibility that the flow will fracture to the surface outside the casing. The potential for cratering is high and the crater could be anywhere. It can be the most destructive when the crater is under the rig or platform. Entire rigs and production platforms have been lost into cratered underground blowouts.

If the casing is set deep, there is the potential for extremely high surface pressures which might result in a failure of the surface equipment or a rupture of the exposed casing strings. If shear rams have been used, there is the potential for even more problems with surface pressures and casing strings.

Underground blowouts are generally more challenging than surface blowouts. The volume of influx is not known nor is the composition. Further, the condition of the wellbore and tubulars which are involved are not reliably descriptive. The well control specialist is confronted with the necessity of analyzing and modeling the blowout and preparing a kill procedure. The tools of analysis and modeling are limited. In addition, the tools and techniques should be limited to only those absolutely necessary since any wireline operation is potentially critical. With the underground blowout, the condition of the wellbore can never be known to a certainty and the risk of sticking or losing wire and tools is significantly increased. Stuck or lost wire and wireline tools can be fatal or at least limit future operational alternatives.

Since the consequences of an underground blowout can be severe, critical questions must be answered:

1. Should the well be shut in?
2. Should the well be vented to the surface?
3. Is the flow fracturing to the surface?
4. Can the losses be confined to a zone underground?
5. Is the casing capable of containing the maximum anticipated pressures?
6. Should the casing annulus be displaced with mud or water?
7. If the casing annulus is to be displaced, what should be the density of the mud?
8. Is the flow endangering the operation and the personnel?

Unlike classical pressure control, there are no solutions which apply to all situations. The underground blowout can normally be analyzed utilizing the surface pressures and temperature surveys. The noise log can be confusing. In all instances, the safety of the personnel working at the surface should be the first concern and the potential for fracturing to the surface must be considered carefully.

It is obviously important to assess the hazards associated with the conditions at the blowout. In temperature survey analysis, it is recognized that the temperature of the flowing fluid will be essentially the same as the temperature of the reservoir from which it came. Therefore, if the flow is from a deep formation into a shallow formation, there should be an abnormally high temperature in the zone of loss.

Surface pressures are a reflection of the conditions downhole. If the surface pressure is high, the zone of loss is deep. Conversely, if the surface pressure is low, the zone of loss is shallow. If the density of the annular fluids is known, the depth to the zone of loss can be calculated.

The noise log is helpful in some instances. The flow of fluids can generally be detected with sensitive listening devices. However, in some instances the blowout was undetected by the noise log while in other instances the interpretation of the noise log indicated the presence of an underground flow when there was none.   The application of these principles is best understood by consideration of specific field examples.

Casing Less than 4,000 feet

With the casing set at less than 4,000 feet, the primary concern is that the underground blowout will fracture to the surface and create a crater. If the blowout is offshore, it is most probable that the crater will occur immediately under the drilling rig. If the productivity is high, then the crater is large and the operation is in great peril. At one operation in the Gulf Coast, several workers were burned to death when the flow fractured to the surface under the rig. At another operation in the Far East, a nine-well platform was lost when the flow fractured to the surface under the platform. At still another operation, a jackup was lost when the crater occurred under one leg. Drill ships have been lost to cratered blowouts.

The blowout at the Pelican Platform offshore Trinidad is a good example. Pelican A-4X was completed at a total depth of 14,235 feet measured depth (13,354 feet true vertical depth). The well was contributing 14 mmscfpd plus 2,200 barrels condensate per day at 2800 psi flowing tubing pressure. Bottomhole pressure was reported to be 5960 psi. The wellbore schematic for the A-4X is shown in Figure 8.1.

At the Pelican A-7, 18 5/8-inch surface casing was set at 1,013 feet and cemented to the surface and drilling operations continued with a 12 ¼-inch hole. The wellbore schematic for the Pelican A-7 is presented as Figure 8.2.

The directional data indicated that the A-7 and A-4X were approximately 10 feet apart at 4,500 feet. However, in the early morning hours, the A-7 inadvertently intercepted the A-4X at 4,583 feet. The bit penetrated the 13 3/8-inch casing, 9 5/8-inch casing and 4 ½-inch tubing. Pressure was lost at the A-4X wellhead on the production deck and the A-7 began to flow. The A-7 was diverted and bridged almost immediately. The A-4X continued to blowout underground. With only 1,013 feet of surface pipe set in the A-7, the entire platform was in danger of being lost if the blowout fractured to the sea floor under the platform.

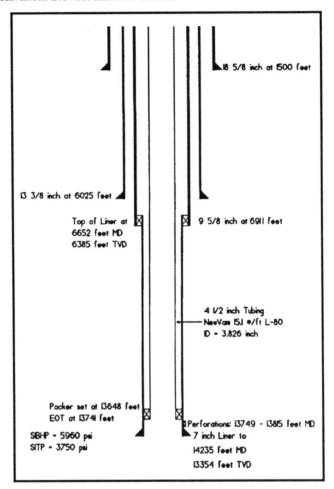

Figure 8.1 - *Pelican A-4X*

After the intercept, the shut-in surface pressure on the tubing annulus at the A-4X stabilized at 2200 psi. The zone into which the production is being lost can be approximated by analyzing this shut-in pressure. The pressure may be analyzed as illustrated in Example 8.1:

Example 8.1
  Given:

  | Fracture gradient, | F_g | = | 0.68 psi/ft |

  | Compressibility factor, | z | = | 0.833 |

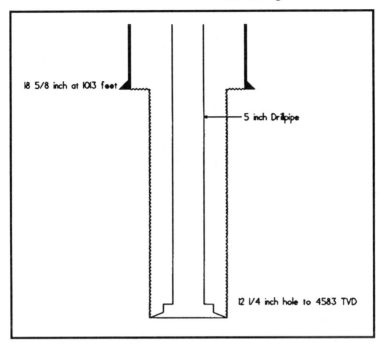

18 5/8 inch at 1013 feet

5 inch Drillpipe

12 1/4 inch hole to 4583 TVD

Figure 8.2 - Pelican A-7

Required:

The depth to the interval being charged

Solution:

The gas gradient, ρ_f, is given by Equation 3.5:

$$\rho_f = \frac{S_g P}{53.3zT}$$

Where:

S_g	=	Specific gravity of the gas
P	=	Pressure, psia
z	=	Compressibility factor
T	=	Temperature, °Rankine

$$\rho_f = \frac{0.6(2215)}{53.3(0.833)(580)}$$

$\rho_f = 0.052$ psi/ft

The depth to the interval being charged is given by Equation 8.1:

$$F_g D = P + \rho_f D \qquad\qquad (8.1)$$

or, rearranging

$$D = \frac{P}{F_g - \rho_f}$$

$$D = \frac{2215}{0.65 - 0.052}$$

$$D = 3,704 \text{ feet}$$

This analysis of the shut-in surface pressure data indicated that the flow was being lost to a zone at approximately 3,700 feet.

As confirmation, a temperature survey was run in the A-7 and is presented as Figure 8.3. Also included in Figure 8.3 are static measurements from the A-3 which were utilized to establish the geothermal gradient. The interpretation of the temperature data was complicated by the fact that the flow path of the hydrocarbons being lost was from the A-4X into the A-7 wellbore and ultimately into a zone in the A-7 wellbore.

The high temperatures at 3,600 feet shown in Figure 8.3 were as expected and consistent with the pressure data analyses indicating the zone of charge to be at approximately 3,700 feet. Pursuant to the analysis of the offset data, the normal temperature at 3,600 feet would be anticipated to be approximately 130 degrees Fahrenheit. However, due to the flow of the gas into the interval at 3,600 feet, the temperature at that zone had been increased 45 degrees to 175 degrees.

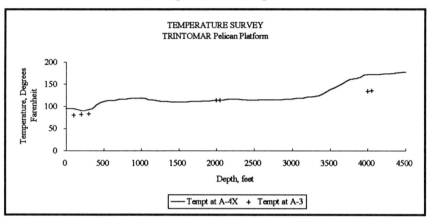

Figure 8.3

By similar analysis, the heating anomaly from 500 to 1,000 feet could be interpreted as charging of sands between 1,000 feet and the sea floor. Pursuant to that interpretation, cratering and loss of the platform could result.

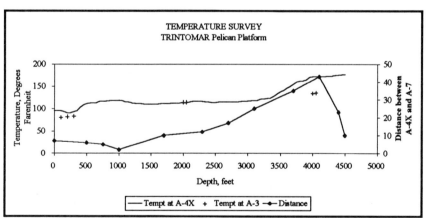

Figure 8.4

Further analysis was warranted. When evaluated in conjunction with the relative position of the two wells, the condition became apparent. Figure 8.4 illustrates the temperature profile in the A-7 overlain by the directional survey analysis of the relative distance between the wellbores.

This figure further confirms the previous pressure and temperature analyses. The two wells are the greatest distance apart, which is 45 feet, at 4,100 feet. That depth corresponds with the most pronounced anomaly, confirming the conclusion that the thermal primary zone of loss is below 3,600 feet. As illustrated, the wellbores are interpreted to be 2 feet apart at 1,000 feet and 5 feet apart at the sea bed. Therefore, the temperature anomaly above 1,000 feet was interpreted to be the result of the proximity of the two wellbores and not caused by the flow of gas and condensate to zones near the sea bed.

Based on the analyses of the surface pressure and the temperature data, it was concluded that working on the platform was not hazardous and that the platform did not have to be abandoned. As further confirmation, no gas or condensate was observed in the sea around the platform at any time during or following the kill operation.

At the Pelican Platform the surface pressures remained constant. When the surface pressures remain constant, the condition of the wellbore is also constant. However, when the surface pressures fail to remain constant, the conditions in the wellbore are, in all probability, changing and causing the changes in the surface pressures.

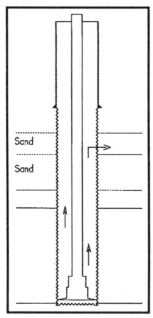

Figure 8.5 - *Offshore Underground Blowout*

Consider an example of an underground blowout at an offshore drilling operation. With only surface casing set, a kick was taken and an underground blowout ensued. The pressures on the drillpipe and annulus stabilized and analysis pursuant to the previous example confirmed that the loss was into sands safely well below the surface casing shoe (Figure 8.5).

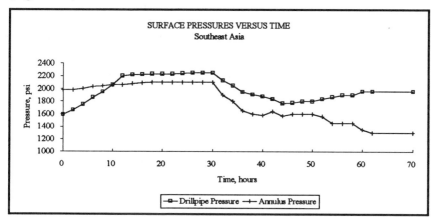

Figure 8.6

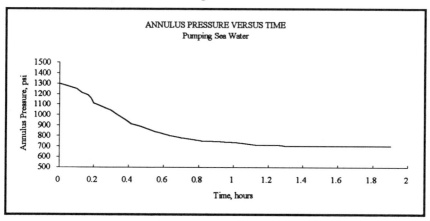

Figure 8.7

The pressure history is presented as Figure 8.6. As illustrated, after remaining essentially constant for approximately 30 hours, both pressures began to change rapidly and dramatically, which was indicative that the conditions in the wellbore were also changing rapidly and dramatically. The pressure changes were confusing and not readily adaptable to analysis and several interpretations were possible. The declining pressures could be indicative that the wellbore was bridging or that the flow was depleting. Further, a change in the composition of the flow could contribute to the change in the pressures. Finally, a decline in annulus pressure could be the result of the flow fracturing toward the surface.

In an effort to define the conditions in the wellbore, a more definitive technique was used to determine precisely the depth of the loss from the underground blowout. With the well shut in, sea water was pumped down the annulus at rates sufficient to displace the gas. As illustrated in Figures 8.6 and 8.7, while pumping, the annulus pressure declined and stabilized. Once the pumps were stopped, the annulus pressure began to increase. With this data, the depth to the loss zone could be determined using Equation 8.2:

$$D = \frac{\Delta P}{\rho_{sw} - \rho_f} \qquad (8.2)$$

Consider Example 8.2:

Example 8.2
   Given:
      Sea water gradient, ρ_{sw} =   0.44 psi/ft

      Gas gradient, ρ_f =   0.04 psi/ft

   Sea water is pumped into the well shown in Figure 8.7 and the surface pressure declines from 1500 to 900 psi as shown in Figure 8.8.

   Required:
      The depth to the zone of loss

   Solution:
      The depth to the zone of loss is given by Equation 8.2:

$$D = \frac{\Delta P}{\rho_{sw} - \rho_f}$$

$$D = \frac{600}{0.44 - 0.04}$$

$D = 1,500 \text{ feet}$

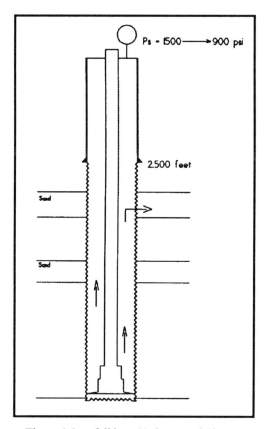

Figure 8.8 - *Offshore Underground Blowout*

As illustrated in Example 8.2, replacing the hydrostatic column of wellbore fluids from the zone of loss to the surface with a hydrostatic column of sea water only reduced the surface pressure by 600 psi. Therefore, the length of the column of sea water between the surface and the zone of loss could be only 1,500 feet which would be 1,000 feet higher than the surface casing shoe. The obvious conclusion would be that the flow was fracturing to the surface. Continuing to work on the location would not be safe. In the actual situation, the flow fractured to the sea floor beneath the rig the following day.

Figure 8.9

In hard rocks the flow may fracture to the surface anywhere. At the Apache Key, shown in Figure 8.9, the well cratered at the wellhead. In the Sahara Desert near the community of Rhourde Nouss, Algeria, the flow cratered a water well 127 meters away from the well (Figure 8.10). It is not uncommon in desert environments for the flow to surface in numerous random locations. It is equally common for the gas to percolate through the sand. At Rhourde Nouss, the hot gas would auto-ignite when it reached the desert floor, producing an eerie blue glow at, and small fires in, the sand (Figure 8.11).

Figure 8.10

Figure 8.11

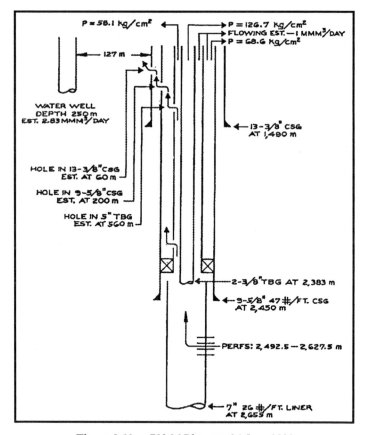

P = 58.1 kg/cm²

127 m

WATER WELL
DEPTH 250 m
EST. 2.83 MMM³/DAY

HOLE IN 13-3/8" CSG
EST. AT 60 m

HOLE IN 9-5/8" CSG
EST. AT 200 m

HOLE IN 5" TBG
EST. AT 560 m

P = 126.7 kg/cm²
FLOWING EST. — 1 MMM³/DAY
P = 68.6 kg/cm²

13-3/8" CSG
AT 1,490 m

2-3/8" TBG AT 2,383 m

9-5/8" 47 #/FT. CSG
AT 2,450 m

PERFS: 2,492.5 — 2,627.5 m

7" 26 #/FT. LINER
AT 2,655 m

Figure 8.12 - *RN-36 Blowout, 24 June 1989*

The wellbore configuration at Rhourde Nouss is illustrated in Figure 8.12. It was critical to know the location of the holes in the tubulars. Since there was flow to the surface, holes had to be present in the 5-inch tubing, 9 5/8-inch casing, and 13 3/8-inch casing. A temperature survey which was run in the 2 3/8-inch tubing which had been run into the well in a kill attempt is illustrated in Figure 8.13. As can be seen, the temperature survey is not definitive. Often a change in temperature or delta temperature can be definitive when the temperature survey alone is not. The delta temperature survey is usually plotted as the change in temperature over a 100-foot interval. The delta survey is then compared with the normal geothermal gradient, which is usually 1.0 to 1.5 degrees per 100 feet. A greater change than normal denotes a problem area. At Rhourde Nouss, when the change in temperature as presented in

Figure 8.14 was analyzed, the tubular failures became apparent. The hole in the 5 ½-inch tubing is the dominant anomaly at 560 meters. The holes in the 9 5/8-inch casing and the 13 3/8-inch casing are defined by the anomaly at 200 meters and 60 meters respectively.

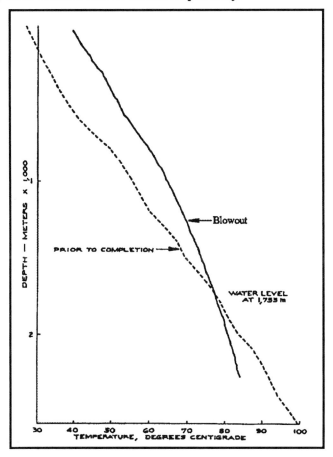

Figure 8.13 - Temperature Survey Comparison

Pipe Below 4,000 feet

With pipe set below 4,000 feet, there is no reported instance of fracturing to the surface from the casing shoe. There are instances of fracturing to the surface after the casing strings have ruptured. Therefore, maximum permissible casing pressures must be established immediately and honored. The maximum annulus pressure at the surface will be the fracture pressure at the shoe less the hydrostatic column of gas.

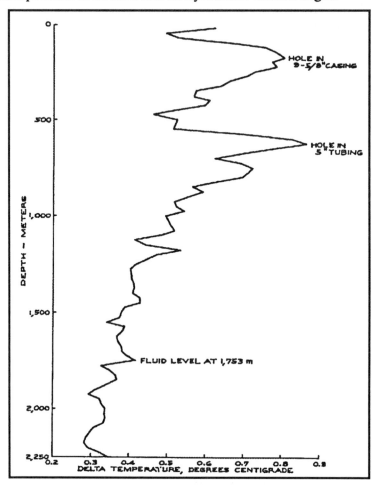

Figure 8.14 - Delta Temperature versus Depth

The wellbore schematic for the Amerada Hess Mil Vid #3 is presented as Figure 8.15. An underground blowout followed a kick at 13,126 feet. The 5-inch drillpipe parted at the 9 5/8-inch casing shoe at 8,730 feet. During the fishing operations which followed, a temperature survey was run inside the fishing string. The temperature survey is presented as Figure 8.16. The 85-degree temperature anomaly at 8,700 feet confirmed the underground blowout. It is interesting to note that the temperature decreased below the top of the drillpipe fish at 8,730 feet.

This anomaly established that the flow was through the drillpipe and that the annulus had bridged.

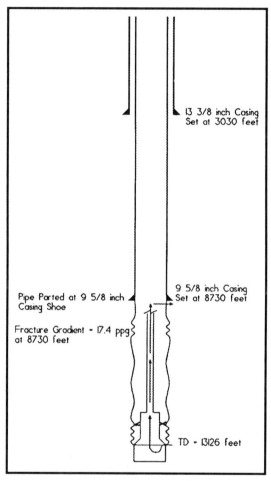

Figure 8.15 - Amerada Hess Mil-Vid #3

It is equally interesting that numerous noise logs run in the same time period failed to detect the underground flow. The noise log run on the Mil Vid #3 is presented as Figure 8.17.

The fracture gradient was measured during drilling to be 0.9 psi/ft. At offset wells the gas gradient was measured to be 0.190 psi/ft. Utilizing this data, the maximum anticipated surface pressure was determined using Equation 8.3:

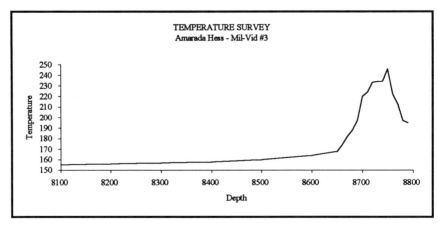

Figure 8.16

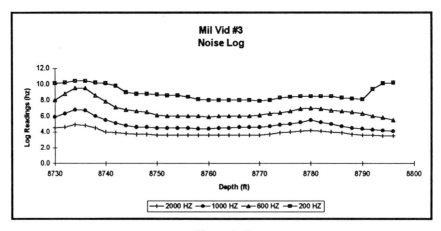

Figure 8.17

$$P_{max} = \left(F_g - \rho_f\right)D \qquad (8.3)$$

$$P_{max} = \left(0.90 - 0.19\right)8730$$

$$P_{max} = \textbf{6198 psi}$$

Although the calculated maximum anticipated surface pressure was only 6200 psi, during subsequent operations the surface pressure was as high as 8000 psi, indicating that the zone of loss was being charged and pressured. Therefore, the actual surface pressure could be much more than the calculated maximum, depending on the volume of flow and the character of the zone of loss.

Once the maximum anticipated surface pressure has been determined, there are three alternatives. The well can remain shut in, provided there is no concern for the integrity of the tubulars. If the pressure is higher than can be tolerated, the well can be vented at the surface, provided that the surface facilities have been properly constructed. Finally, mud or water can be pumped down the annulus to maintain the pressure at acceptable values.

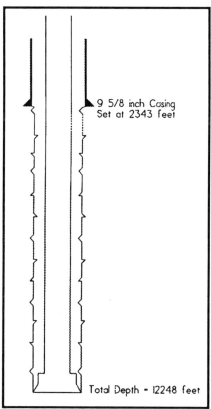

9 5/8 inch Casing
Set at 2343 feet

Total Depth - 12248 feet

Figure 8.18 - *Thermal Exploration - Sagebrush 42-26*
Sweetwater County, Wyoming

Since the anticipated surface pressures were unacceptable and the Mil-Vid #3 was located within the town of Vidor, Texas, the only alternative was to pump mud continuously into the drillpipe annulus. Accordingly, 16-ppg to 20-ppg mud was pumped continuously down the drillpipe annulus for more than 30 days — an expensive but necessary operation. However, the surface pressures were maintained below an acceptable 1000 psi.

As a final comparison between the noise log interpretation and the temperature survey, consider the well control problem at the Thermal Exploration Sagebrush No. 42-26 located in Sweetwater County, Wyoming. The wellbore schematic is presented as Figure 8.18. During drilling at approximately 12,230 feet, a kick was taken, the well was shut in and an underground blowout ensued. Water flows from intervals above 4,000 feet further complicated analysis.

In an effort to understand the problem, a temperature survey was run and is presented as Figure 8.19. As illustrated in Figure 8.19, the temperature gradient between the top of the drill collars at 11,700 feet and 5,570 feet was normal at 1.25 degrees per 100 feet. With a normal gradient, it is conclusive that there can be no flow from the interval at 12,230 feet to any interval in the hole. The significant drop in temperature at 5,570 feet indicated that the well was flowing from this depth or that a lost circulation zone was at this depth. The temperature survey was conclusive that the well was flowing above 4,000 feet since the gradient above this point is essentially 0.

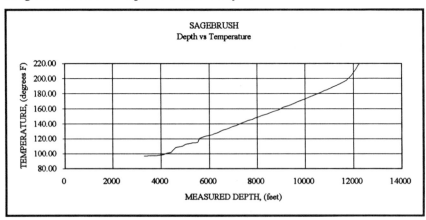

Figure 8.19

The noise log is presented as Figure 8.20. As illustrated in Figure 8.20, noise anomalies are indicated at 4,000 feet and 6,000 feet, which correspond to the temperature interpretation that fluid was moving at these depths. However, in addition, the noise log was interpreted to indicate flow from 12,000 feet to 7,500 feet, which was in conflict with the interpretation of the temperature survey. Subsequent operations proved conclusively that there was no flow from the bottom of the hole at the time that these logs were run.

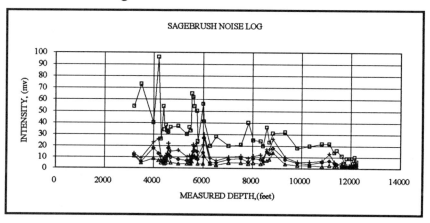

Figure 8.20

Charged Intervals – Close Order Seismic – Vent Wells

In underground blowouts, the charging of the zone of loss is an important consideration for relief well operations. A relief well located within the charged interval will encounter the charged interval and experience well control problems. The mud weight required to control the charged interval can approach 1 psi/ft, which usually exceeds the fracture gradient of the intervals immediately above and below the zone of loss. Therefore, the relief well can be lost or another casing string can be necessitated.

In addition, the charging of the zone of loss is an important consideration in analyzing the potential for the influx to fracture to the surface. Nowhere is the question of shallow charging more important than offshore.

In late September, 1984, Mobil experienced a major blowout at its N-91 West Venture gas field, offshore Nova Scotia, Canada. A relief well, the B-92, was spudded approximately 3,000 feet from the blowout. During drilling at 2,350 feet with conductor set at 635 feet, a gas kick was taken. The gas zone encountered was the result of the charging caused by the N-91 blowout. A shallow seismic survey was conducted to assist in defining the extent of the underground charging. Booth reported that, when the seismic data was compared with the original work, two new seismic events were identified.[1] The deeper event occurred at about 2,200 to 2,300 feet, which corresponded to the charged zone in the relief well. However, there was also a second event at approximately 1,370 to 1,480 feet. The upper interval was interpreted to be approximately 3,300 feet in diameter emanating from the N-91. This event was of great concern since only unconsolidated sandstones, gravels and clays were present between the charged interval and the ocean floor 1,100 feet away.

Fortunately, the charged interval never fractured to the surface. Eight additional surveys were conducted between 5 November, 1984 and 9 May, 1985. Those surveys revealed that the gas in the shallow zone had not grown significantly since the first survey and had migrated only slightly up dip. In addition, the surveys were vital for the selection of safe areas for relief well operations. Finally, the surveys were vital in analyzing the safety and potential hazard of continuing operations onboard the Zapata Scotain with the rig on the blowout.

In the past, it has been customary to drill vent wells into the charged zones in an effort to reduce the charging. Generally, such efforts have not proven successful. The zones of loss are normally not good quality reservoir. Therefore, the amount of gas being lost greatly exceeds that recovered from the vent wells. The result is that the charging is relatively unaffected by the vent wells.

At the TXO Marshall, for example, three vent wells were completed. The blowout was discovered to be losing approximately 15 mmscfpd underground. The three vent wells were producing a total of less than 2 mmscfpd. Experiences such as this are commonly reported.

If charging is a problem, the better alternative may be to vent the blowout at the surface. If charging is to be affected, the volume of gas vented would have to be sufficient to cause the flowing surface pressure to be less than the shut-in surface pressure plus the frictional losses between the zone of loss and the surface. Once the charging is stopped, the operations at the surface can be conducted safely and the relief well, if necessary, can be in the most expeditious position.

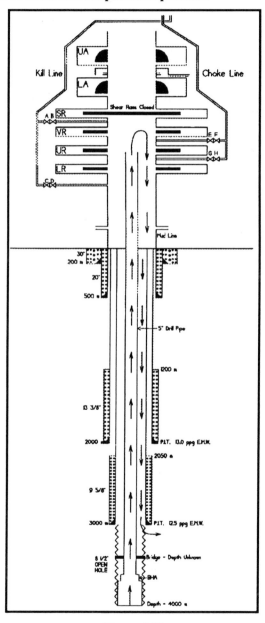

Figurre 8.21

Shear Rams

If shear rams have been used, the situation is very similar to that at the Mil-Vid #3 in that the drillpipe has been severed immediately below the shear rams and the flow is usually through the drillpipe, down the drillpipe annulus and into the formations below the shoe.

A typical example is illustrated in Figure 8.21. As illustrated, when the shear rams are used, the result is often very similar to setting pipe and completing for production. If flow is only through the annulus, the well will often bridge, especially offshore in younger rocks which are more unstable. With pipe set and open to the shoe, the flow can continue indefinitely.

Cement and Barite Plugs

Generally, when an underground blowout occurs, the impulse is to start pumping cement or setting barite pill. Cement can cause terminal damage to the well. At least, such indiscriminate actions can result in deterioration of the condition of the well. It is usually preferable to bring the blowout under control prior to pumping cement. If the problem is not solved, when the cement sets, access to the blowout interval can be lost. Cement should not be considered under most circumstances until the well is under control or in those instances when it is certain that the cement will control the well.

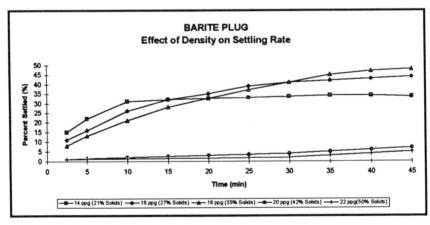

Figure 8.22

Barite pills can be fatal or cause the condition of the wellbore to deteriorate. A barite pill is simply barite, water and thinner mixed to approximately 18 ppg. Each mixture can demonstrate different properties and should be pilot tested to insure proper settling. Intuitively, the heavier, the better for a barite pill. However, consider Figure 8.22. As illustrated, for this particular mixture, barite pills ranging in weight from 14 ppg to 18 ppg would permit the barite to settle. However, mixtures above 20 ppg failed to settle barite. The failure to settle is caused by the interaction of the barite particles. All mixtures can attain a density which will not permit the settling of barite. When setting a barite pill, the total hydrostatic does not have to exceed the reservoir pressure for the barite pill to be successful. However, the greatest success is experienced when the total hydrostatic does exceed the reservoir pressure.

When barite or cement is chosen, the drill string should be sacrificed. Attempting to pull the drill string out of a cement plug or barite pill only retards the setting of the cement or the settling of the barite.

Too often, cement plugs and barite pills only complicate well control problems. For example, in one instance in East Texas, the improper use of cement resulted in the loss of the well and a relief well had to be drilled. In another instance offshore, the barite pill settled on the drillpipe as it was being pulled and the drillpipe parted, causing loss of the well. In both instances, millions of dollars were lost · due to the unfortunate selection of barite and cement.

Reference

1.   Booth, Jake, "Use of Shallow Seismic Data in Relief Well Planning," <u>World Oil</u>, May 1990, page 39.

EPILOGUE
The AL-AWDA Project
The Oil Fires of Kuwait

No text on advanced pressure control would be complete without a brief history and overview of this historical project. I am proud to have served the Kuwait Oil Company and the Kuwaiti people in their effort. I consider my involvement one of the greatest honors of my career.

No picture can capture — no language can verbalize — the majesty of the project. It was indeed beyond description by those present and beyond complete appreciation by those not present. A typical scene is shown in Figure 1. All were photographed during the day. The smoke turned day to night.

Figure 1

The rape of Kuwait was complete. The retreating Iraqi troops had savagely destroyed everything in the oil fields. There was nothing left to work with. There were no hand tools, no pump trucks, no cars, no pickups, no housing — nothing. Everything necessary to accomplish the goal of extinguishing the fires and capping the wells had to be imported. The world owes the valiant Kuwaiti's a great debt. To extinguish almost 700 oil well fires in eight months is an incredible accomplishment, especially in light of the fact that Kuwait is a very small country of only 1.5 million people and had been completely and ruthlessly pillaged. No one worked harder or longer days than the Kuwaiti's. Many did not see their families for months and worked day after day from early morning until well after dark for days, weeks and months in their tireless effort to save their country. They did an incredible job.

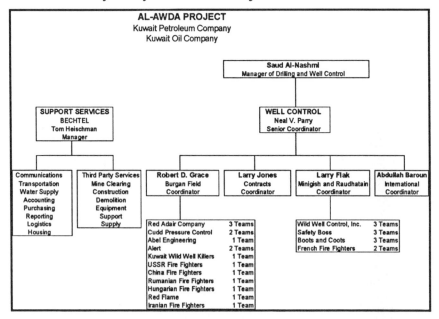

Figure 2

By early summer 1991, Kuwait Oil Company was displeased with the progress of the project. As a result, Mr. Neal V. Parry, former president of Santa Fe Drilling Company, was asked by Kuwait Oil Company to serve as Senior Coordinator for Fire Fighting and assumed those responsibilities on 1 August. Mr. Parry reported to Mr. Saud Al-Nashmi, Manager of Drilling and Well Control for Kuwait Oil Company. As understood by the author, the basic organization chart, effective after 1 August 1991, as pertained to fire fighting and well control, is presented as Figure 2. The oil fields of Kuwait are shown in Figure 3. Greater Burgan Field is the largest oil field in Kuwait. Mr. Larry Flak was the coordinator for the fields outside Burgan, which were Minigish and Raudatain. Texaco was responsible for the wells in the neutral zone and the British Consortium was responsible for Sabriyah. The Kuwaiti Wild Well Killers, or KWWK, were responsible for the wells in Umm Gudair. Mr. Larry Jones, also a former Santa Fe employee, was charged with contracts and logistics. Mr. Abdoulla Baroun, an employee of Kuwait Oil Company, liaisoned between Kuwait Oil Company and the multinational teams.

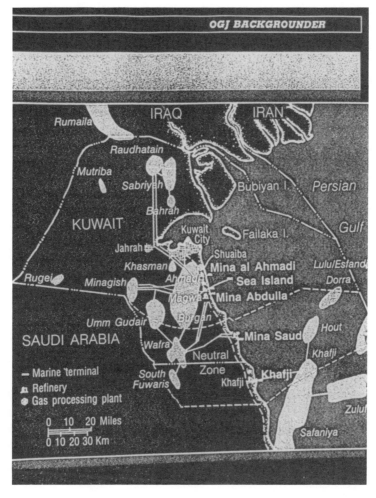

Figure 3

Until 1 August, eight to twelve teams from four companies had controlled 257 wells with most of the wells being in the Ahmadi and Magwa Fields, which are nearest to Kuwait City. After Magwa and Ahmadi, the primary emphasis was on the Burgan Field. However, as additional teams arrived, the original teams were moved to the fields outside Burgan. By the end of the project in early November, there were twenty-seven fire fighting teams deployed in Kuwait as shown in the organization structure of Figure 2.

Thousands were involved in these critical operations and all deserve mention. Almost all of the support was provided by Bechtel under the very capable management of Mr. Tom Heischman. Texaco furnished support in the Neutral Zone, and the British Consortium furnished most of the support in Sabriyah. A substantial contribution was made by the management and employees of Santa Fe Drilling Company; many of whom were among the first to return to Kuwait after the war. One of the many contributions made by Santa Fe Drilling Company was the supply of heavy equipment operators who worked side-by-side with fire fighters to clear the debris and extinguish the fires.

An early report of the status of the fields and wells in Kuwait is presented as Table 1. In most of the fields, it was easy to determine the status of each individual well. Such was not the case in Burgan Field. In Burgan, the well density is very high. The smoke reduced visibility to a few feet, and access to some parts of the field was impossible until the very end of the project. Even in the last few weeks, there was disagreement concerning the status of individual wells. However, the totals were very accurate considering the circumstances. A typical day in Burgan Field is shown in Figure 1.

The majority of the wells in Kuwait are older and shallow (less than 5,000 feet) with surface pressures less than 1000 psi. Typically, they were completed with 3 ½-inch tubing inside 7-inch casing and produced through both the casing and the tubing. The older wells had the old style Gray Compact Head that houses all of the casing hangers in one body in progressively larger mandrels. A Gray Compact Head is pictured in Figure 4. The newer and deeper wells had higher pressures and more conventional wellheads.

The Iraqi troops packed plastic explosives around the bottom master valve on the tree as well as the wing valves on the "B" section.

Sand bags were then packed on top of the explosives to force the explosion into the tree. The force of the explosion was tremendous and the damage indescribable.

The AL-AWDA Project
Oil Wells Survey Data
by Field

Field	Drilled	On Fire	Oil Spray	Damaged	Intact	Remarks
Magwa	148	99	5	21	15	
Ahmadi	89	60	3	17	6	
Burgan	423	291	24	27	67	
Raudhatain	84	62	3	5	3	
Sabriyah	71	39	4	9	2	
Ratqa	114	0	0	0	8	Figure excludes shallow wells
Bahra	9	3	2	*	*	* Denotes uncertain about figure
Minagish	39	27	0	7	1	
Umm Gudair	44	26	2	10	2	
Dharif	4	0	0	0	3	
Abduliyah	5	0	0	0	4	
Khashman	7	0	0	1	1	
Total	1037	607	43	97	112	

Table 1

Figure 4

In fortunate instances, such as pictured in Figure 5, the tree was blown off at the bottom master valve and there was no other damage. In that case, the fire burned straight up and the oil was almost totally consumed. However, the vast majority of cases were not as fortunate. The destruction of the tree was not complete and, as a result, oil flowed out multiple cracks and breaks. The consequence was that the combustion process was incomplete. The unburned oil collected around the wellhead in lakes that were often several feet deep (Figure 6). The ground throughout the Burgan Field was covered with several inches of oil. Ground fires covering hundreds of acres were everywhere (Figure 7). In addition, some of the escaping unburned oil was cooked at the wellhead and formed giant mounds of coke (Figure 8).

Figure 5

While the coke mounds were grief to the fire fighters, perhaps they were a benefit to Kuwait and the world. The coke accumulations served to choke the flow from the affected wells. In all probability, the reduction in flow rate more than offset the additional time required to cap the well.

Figure 6

Figure 7

Figure 8

THE PROBLEMS

The Wind

The wind in Kuwait was a severe problem. Normally, it was strong from the north to northwest. The strong, consistent wind was an asset to the fire fighting operation. However, during the summer months, the hot wind was extremely unpleasant. In addition, the sand carried by the wind severely irritated the eyes. The only good protection for the eyes was to wear ordinary ski goggles. It was expected that the summer "schmals" or wind storms would significantly delay the operations. But such was not the case. The oil spilled onto the desert served to hold the sand and minimize the intensity of the wind storms. As a result, the operation suffered few delays due to sand storms.

The wind was most problematic on those occasions when there was no wind. During these periods, it was not uncommon for the wind direction to change 180 degrees within fifteen minutes. In addition, the wind direction would continue to change. Any equipment near the well might be caught and destroyed by the wind change. In any event, all of the equipment would be covered with oil, and the operation would be delayed until it could be cleaned sufficiently to continue. These conditions could persist for several days before the wind once again shifted to the traditional northerly direction.

The humidity in the desert was normally very low. However, when the wind shifted and brought the moist Gulf air inland, the humidity would increase to nearly 100%. When that happened, the road would become very slick and dangerous. On several occasions there were serious accidents. In one case, a man was paralyzed as a result of an automobile accident caused by the slick road.

Logistics

The first problem was to get to the location. Access was provided to the fire fighters by the EOD who cleared the area of explosives remaining after the war and by Bechtel who was responsible for furnishing the location and supplying water for fire fighting. Everyone involved in this aspect (and there were many) did an incredible job. Close cooperation was vital in order to maintain efficiency in strategy planning. The goal was to keep all teams working to the very end of the project.

During the height of the activity in September and October, it was not unusual to haul 1,500 dump truck loads of road and location building material each day and several hundred loads during the night. The Filipino truck drivers were not supposed to get close to the fires as they hauled to the locations; however, it was common to find a dump truck with blistered paint.

It was not possible to survey the locations in the oil lakes for munitions. Therefore, access was safely gained by backing the truck to the end of the road and dumping a few cubic yards of material into the lake, hoping that the dirt would cover any ordinance. A dozer would then spread the material and the process would be repeated until the location was reached.

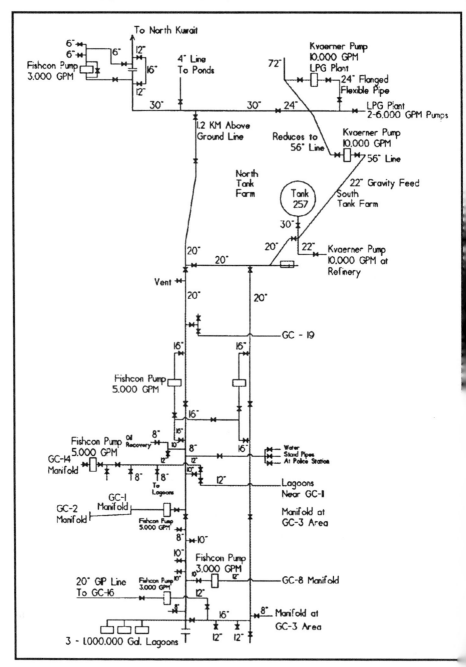

Figure 9 - *The Water Supply System*

Water

The water system was constructed from the old oil gathering lines (illustrated in Figure 9). Most of the lines had been in the ground for many years and all had been subjected to munitions. Therefore, it was not unusual for a water line laid across the desert to look like a sprinkler system. In spite of everything, 25 million gallons of water were moved each day. Lagoons were dug and lined at each location, when the gathering system was capable of supporting the lagoon with water. The capacity of the lagoons was approximately 25,000 barrels. Sufficient water was always a problem. Water was continually pumped into the lagoon during the fire fighting phase of the operation. Generally, the lagoon would fill overnight. During the last days of the project, two lagoons were constructed in the Burgan Field at locations predicted to be difficult. Because the team concentration increased near the end, the demand on the isolated working area was very great. In spite of all obstacles, in these last days, water was not a problem for the first time during the project. In areas where water could not be transported by pipe line, frac tanks were used and water was trucked from nearby loading points. Randy Cross, a most pleasant and capable New Zealander employed by Bechtel, was primarily responsible for water and logistics in Burgan.

Ground Fires

Once the location was reached, the fire fighters took over and spread the material to the well. In the process, the ground fires had to be controlled and were a major problem. The ground fires often covered tens of acres. In many instances, we were not able to identify the well. The ground fires were fed by the unburned oil flowing from the coke mounds or from the wells themselves after the fire had been extinguished. Most of the time, the wild well fighters worked on a live well with a ground fire burning less than 100 yards down wind.

The worst ground fires were in the heart of Burgan. In anticipation of the problems, together with Safety Boss personnel, a unit was specially designed to fight the ground fires. It consisted of a 250-bbl tank mounted on Athey Wagon tracks. A fire monitor was mounted on top of the tank and a fire pump was mounted on the rear of the tank. It was pulled by a D-9 Cat and followed by a D-8 for safety. It had a built-in inductor tube to disperse various foaming agents commonly used in fire fighting. The crew routinely worked between the blowing well and the ground fire and would be covered with oil every day. "Foamy One," pictured in Figure 10, and her crew made a tremendous difference and contribution to the effort.

Figure 10

The Rumanian and Russian fire fighters were particularly good at fighting and controlling ground fires. They both brought fire trucks capable of spreading powder and chemicals to smother the ground fires. Their efforts significantly contributed to the success of the project. In all cases, the hot spots had to be covered with dirt to prevent re-ignition. Some of the hot spots continued to burn for months. They often were not visible during the day. However, at night the desert fire was clearly visible. One of the last projects was to cover these persistent devils.

Oil Lakes

In addition to the ground fires, unburned oil had gathered into huge lakes. Often the lake surrounded the well. In one instance, fire came out one side of the coke pile and a river of oil flowed down the other side of the coke pile. The lakes could be several feet deep. Often the lakes caught fire and burned with unimaginable intensity, producing tremendous volumes of smoke. Working in the lakes was very dangerous because the access roads could become bounded by fire, trapping the workers. After the oil weathered for several days, it was less likely to burn. Therefore, most lake fires could be extinguished by eliminating the source of fresh oil. In the latter days, the problem was solved by surrounding the burning well in the lake with a road approximately 50 feet in width. The road was then crossed by fire jumps in strategic locations that isolated the fire from the fresh oil in the lake. The approach was successful. Had some of the bigger lakes caught fire, they would have burned for weeks.

Figure 11

The Coke Piles

Once the well was reached, the wellhead had to be accessed. Typically, each fire fighting team had fire pumps, Athey Wagons, monitor sheds, cranes, and two backhoes — a long reach and a Caterpillar 235. The first step was to spray water on the fire from the monitor sheds in order to get close enough to work. The average fire pump was capable of pumping approximately 100 barrels of water per minute and two were usually rigged up on each well. Although the lagoons contained approximately 25,000 barrels of water, they could be depleted very rapidly at 200 barrels per minute. Using the water cover, the monitor sheds could be moved to within 50 feet of the wellhead. The long reach backhoe could then be used to dig away the coke pile and expose the well head. This operation is illustrated in Figure 11.

As previously mentioned, some of the unburned oil cooked around the wellhead to form a giant coke pile. The coke pile formed like a pancake 100 feet in diameter. At the wellhead itself, the coke pile might be as large as 30 to 40 feet high and 50 to 70 feet in diameter. It had the appearance of butter on top of a pancake. In some instances the coke piles were very hard and difficult to dig. In other instances, the coke was porous and easily removed. It was not unusual for fire to burn out one side of the coke pile while oil flowed out another side. In the northern fields, berms had been constructed around the wellheads. These quickly filled up with coke. Digging the coke resulted in a pot of boiling, burning oil.

Summary of Kill Techniques	
Stinger	225
Kill Spool	239
Capping Stack	94
Packer	11
Other	121
Total	690

Table 2

CONTROL PROCEDURES

After exposing the wellhead, the damage was assessed to determine the kill procedure. Eighty-one percent of the wells in Kuwait were controlled in one of three procedures. The exact proportions are presented in Table 2.

The Stinger

If the well flowed straight up through something reasonably round in shape, it would be controlled using a stinger. As shown in Table 2, a total of 225 wells were controlled using this technique. A stinger was simply a tapered sub that was forced into the opening while the well flowed and sometimes while it was still on fire. The stinger was attached to the end of a crane or Athey Wagon. The kill mud was then pumped through the stinger and into the well. If the opening was irregular, materials of irregular shapes and sizes were pumped to seal around the stinger. Due to the low flowing surface pressures of most of the wells, the stinger operations were successful. However, stingers were not normally successful on openings larger than seven inches or on higher pressured wells. A typical stinger operation is schematically illustrated in Figure 12.

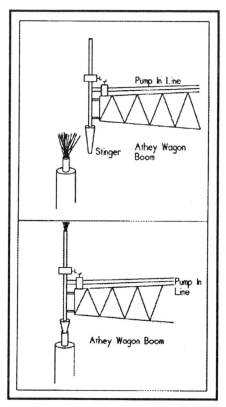

Figure 12

The Capping Spool

Another popular alternative was to strip the wellhead to the first usable flange. Using a crane or an Athey Wagon, a spool with a large ball valve on top was then snubbed onto the usable flange remaining on the tree. The valve was then closed and the well killed through a side outlet on the spool below the ball valve. This procedure was referred to as a capping spool operation and is illustrated in Figures 13 and 14. As shown in Table 2, 239 wells were controlled using this technique. This operation was performed after the fire had been extinguished.

Figure 13

The Capping Stack

Failing the aforementioned alternatives, the wellhead was completely stripped from the casing. This operation was accomplished with and without extinguishing the fire. In some instances, the wellhead was pulled off with the Athey Wagon. In other instances, it was blown off with explosives. In the early days of the effort, it was cut off using swabbing units and wirelines. In the final four months, the procedure often involved the use of high pressure water jet cutters. (Cutting will be discussed later.) After the wellhead was removed, the casing strings were stripped off using mechanical cutters, commonly known as port-a-lathes, leaving approximately 4 feet of the string to receive the capping stack.

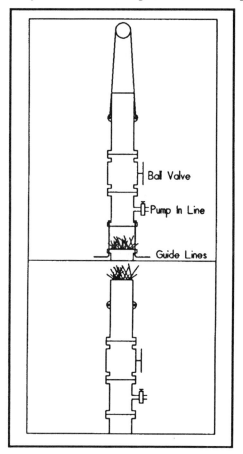

Figure 14 - Kill Spool

The wells were normally capped with a capping stack on the 7-inch production casing. Fortunately, all of the casing strings in the wells in Kuwait were cemented to the surface. Therefore, when the wellheads were severed from the casing, they did not drop or self destruct as would commonly occur in most wells. The capping stack consisted of three sets of blowout preventers (Figure 15). The first set were slip rams designed to resist the upward force caused by the shut-in well. The second set were pipe rams turned upside down in order to seal on the exposed 7-inch casing.   A spool with side outlets separated the slip rams from the uppermost blind rams.

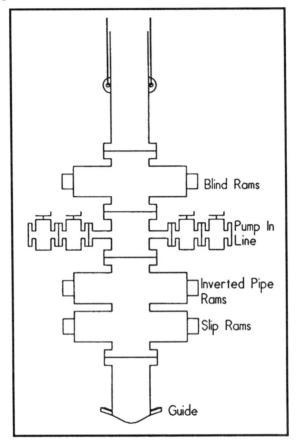

Figure 15 - Capping Stack

After the wellhead was removed, the fire was extinguished.   The casing was stripped, leaving approximately 4 feet of 7-inch casing exposed.   Normally, a crane would be used to place the capping stack on the exposed casing.   Once the capping stack was in place, the pipe rams were closed, followed by the slip rams.   The well would now flow through the capping stack.   When the pump trucks were connected and all was ready, the blind rams were closed.   The pump trucks then pumped into the well.   This operation has been performed elsewhere on burning wells, but was performed in Kuwait only after the fires were extinguished.   The capping stack operation was performed on 94 wells in Kuwait. BG-376 was the last blowout in the Burgan Field and was controlled on 2 November 1991. Figure 16 pictures the capping stack operation at BG-376.

Figure 16

EXTINGUISHING THE FIRES

Water

Extinguishing the fires in Kuwait was the easiest part of the entire operation. Three basic procedures were employed, though no official records were kept. However, the vast majority of the fires were extinguished using the water monitors. The oil contained asphalt and was of low gravity; therefore, it was less volatile than most crudes. Usually three to five fire monitors were moved close to the fire and the flow was intensified at the base of the fire. As the area cooled, the fire began to be interrupted. One monitor sprayed up along the plume to further cool the fire. Very large well fires were extinguished in just minutes. Figure 17 illustrates the use of the fire monitors to extinguish a fire.

Figure 17

Some of the teams used fire suppressant materials and chemicals to effectively extinguish the fires and minimize their water requirements. A much broader use of these materials would have more effectively conserved the precious water supplies.

Nitrogen

Prior to the availability of the required volumes of water, many fires were extinguished using nitrogen. A 40-foot chimney attached to the end of an Athey Wagon was placed over the fire, causing the flow to be directed up through the chimney. The fire then burned only out of the top of the chimney. Nitrogen was then injected into the chimney through an inlet at the base. A typical chimney is shown in Figure 17.

Explosives

Of course, some of the fires were extinguished using explosives. Explosives effectively rob the fire of the oxygen necessary to support combustion. The fire monitors were used to cool the fire and the area around the fire in order to prevent re-ignition. Charge size ranged from five pounds of C_4 to 400 pounds of dynamite. The charges were packed into a drum attached to the end of an Athey Wagon boom. Some included fire suppressing materials along with the explosives. The drum was wrapped with insulating material to assure that the explosives did not merely burn up in the fire. The explosives were then positioned at the base of the plume and detonated. Figure 18 illustrates a charge being positioned for detonation.

Figure 18

Novel Techniques

One technique captured considerable publicity and interest. The Eastern Bloc countries — Russia, Hungary, and Rumania — used jet engines to extinguish the fire. The Hungarian fire fighters were the most interesting. Their "Big Wind" consisted of two MIG engines mounted on a 1950's vintage Russian tank. Water and fire suppressants were injected through nozzles and into the vortex by remote control. The tank was positioned approximately 75 feet from the fire and then the water lines were connected. The engines were turned on at low speed and the water started to protect the machine as it approached the fire. The tank was then backed toward the fire. Once positioned, the speed of the engines was increased and the fire was literally blown out as one would blow out a match. The Hungarian "Big Wind" is illustrated in Figure 19.

Figure 19

CUTTING

In the early days of the fire fighting operation, a steel line between two swab units was used to saw casing strings and wellheads. This technique proved to be too slow. By early August, pneumatic jet cutting had completely replaced the swab lines. Water jet cutting is not new technology even to the well control business; however, techniques were improved in Kuwait.

Primarily two systems were used in Kuwait. The most widely used was the 36000 psi high pressure jet system using garnet sand. A small jet was used with water at three to four gallons per minute. Most often, the cutter ran on a track around the object to be cut. In other instances, a hand-held gun was used, which proved to be very effective and useful. More than 400 wells employed this technique. One limitation was that the fire had to be extinguished prior to the cut, and the cellar prepared for men to work safely at the wellhead. Another aspect had to be evaluated before widespread use was recommended. In the dark of the smoky skies or late in the evening, the garnet sand caused sparks as it impacted the object being cut. It was not known if under the proper circumstances these sparks would have been sufficient to ignite the flow. However, re-ignition was always of concern.

Another water jet system used a 3/16 jet on a trac or yoke attached to the end of an Athey Wagon boom. The trac permitted cutting from one side with one jet while the yoke involved two jets and cut the object from both sides. The system operated at pressures ranging between 7500 psi and 12500 psi. Gelled water with sand concentrations between 1 and 2 ppg were used to cut. The system was effective and did not require the men to be near the wellhead. In addition, the system could be used on burning wells, provided the object to be cut could be seen through the fire. This technology was used on 48 wells.

Conventional cutting torches were used by some. A chimney was used to elevate the fire and the workers would cut around the wellhead. Magnesium rods were also used because they offered the advantage of coming in 10-foot lengths that could be telescoped together.

STATISTICS

The best authorities predicted that the operation would require five years. It required 229 days. The project's progress is illustrated in Figures 20 and 21. Originally, there were only four companies involved in the fire fighting effort. At the beginning of August, additional teams were added, making a total maximum of twenty-seven teams from all over the world. The companies that participated and the number of wells controlled by each company are shown in Table 3. The number of wells controlled by each company listed in Table 3 is not significant because some companies had more crews and were in Kuwait for a longer period of time. What is significant is that the most difficult wells were controlled after 1 August. As shown in Figure 21, the number of team days per well was essentially constant at approximately four team days per well. That is not to say that some wells were not more difficult and that some teams were not better.

The Al-AWDA Project
OIL WELLS SECURED AND CAPPED

CONTRACTOR'S NAME		TOTAL
RED ADAIR (American)		111
BOOTS & COOTS (American)		126
WILD WELL CONTROL (American)		120
SAFETY BOSS (Canadian)		176
CUDD PRESSURE CONTROL (American)		23
NIOC (Iran)		20
CHINA PETROLEUM		10
KUWAIT OIL COMPANY (KOC)		41
ALERT DISASTER (Canadian)		11
HUNGARIAN		9
ABEL ENGINEERING (American) -KOC		8
	-WAFRA	31
RUMANIAN (Romania)		6
RED FLAME (Canadian)	-KOC	2
	-WAFRA	5
HORWELL (French)		9
RUSSIAN		4
BRITISH		6
PRODUCTION MAINTENANCE		9
TOTAL		727

Table 3

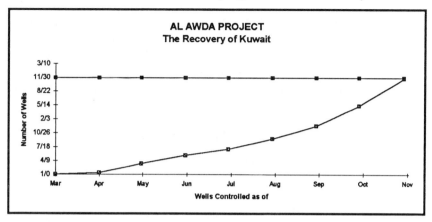

Figure 20

As also illustrated in Figure 21, the high was reached in the week ending 12 October, when a record fifty-four wells were controlled in seven days. For the month of October, 195 wells were controlled for an average of 6.3 wells per day. The record number of wells controlled in a single day was thirteen.

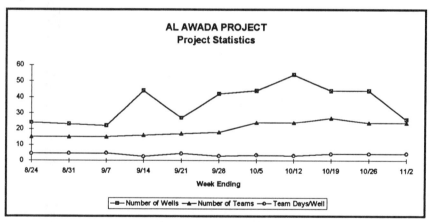

Figure 21

Figures 22 and 23 illustrate the statistics for the Burgan Field. A high of fifteen teams were working in Burgan in late August. Beginning in early September, the teams were moved to the fields in the north and west. They were replaced with teams from around the world. Figures 21 and 23 show the number of team days per well actually decreased between August and November from an average of five team days per well to an

average of four team days per well. As illustrated in Figure 23, the
progress in Burgan was consistent. As can be noted in Figure 23, the best
progress was recorded during the week ending 14 September when
twenty-nine wells were controlled by ten teams at just over two team days
per well. The tougher wells in Burgan were controlled in the latter days of
the project.

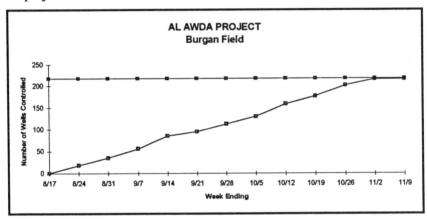

Figure 22

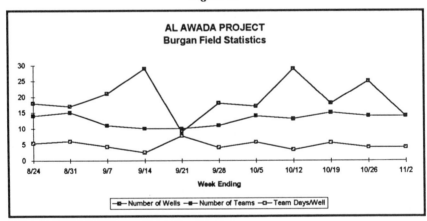

Figure 23

SAFETY

It is incredible that 696 wells were extinguished, capped, and killed over an eight-month period with no fatalities at the wellhead. There were ten fatalities associated with the Kuwait fires as of the official end of the project on 6 November 1991.

Every fire fighting team was assigned an ambulance and a medic. These men were life savers in this harsh environment. They were ever present with water and other drinks for the parched workers. In addition, they were able to treat minor illnesses and provide support for medical helicopters. It was routine to have an injured man in the hospital within ten to fifteen minutes of an injury being reported.

The British Royal Ordinance did a marvelous job of clearing the well sites of mines and other munitions. A path and area for the fire fighters was routinely checked. Any explosives too near the flames would have been consumed by the fire. Therefore, the fire fighters were safe from these problems. However, by the end of the project, EOD had lost two men to land mine clearance. One was killed while clearing a beach. The other was killed while working a mine field in the Umm Gudair Field when the mine inexplicably detonated.

The worst accident occurred very early in the project. Smoke from a burning well along the main road between Ahmadi and Burgan routinely drifted across the road. Just as routinely, the crews became accustomed to the situation and regularly drove through the smoke. On that fatal day, there was fire in the smoke and in the ditches beside the road. Four service company men in one vehicle and two journalists in another vehicle apparently became disoriented in the smoke and drove into the fire. All six perished.

In other instances, one man was killed in a pipeline accident and another was killed in a road accident involving heavy equipment.

Of the fire fighters, the Chinese team had one man severely burned, but his life was never in danger. In addition, at the end of the project, a member of the Rumanian team was badly burned late one evening when the wind died and gas fumes gathered. The fumes were ignited by a hot spot from a previously extinguished ground fire. It was thought that he would not survive because he had inhaled the fire. However, he was air lifted to Europe and has recovered.

CONCLUSION

At the official end of the Al-Awda Project on 6 November 1991, there were 696 wells on the report. The first well was secured on 22 March and the last fire was extinguished on 6 November for a total of 229 days. Remarkable!

INDEX